DIMENSIONAL
THROUGH PERSPECTIVE

A Reference Manual

James R. Williamson
Michael H. Brill

KENDALL/HUNT PUBLISHING COMPANY
2460 Kerper Boulevard P.O. Box 539 Dubuque, Iowa 52004-0539

Table of Contents

Table of Contents

— Graphical Chapters —

Table of Contents

Table of Contents

Table of Contents

Table of Contents

Appendix:

List of Figures

List of Figures

List of Tables

PREFACE

In the classical sense of photogrammetry, perspective is one of the first geometric properties of photography that is studied. The emphasis of such studies is usually on aerial photography, where the aerial photographs may be vertical or low or high oblique views. Most courses of instruction quickly replace single photographs by multiple photograph studies because of the ease with which multiple photographs can be measured, computer programs run, and data analyzed. It is rare that such emphasis is put into the discussion of single-photograph perspective required for close-range forensic photogrammetry. It is the purpose of this manual to provide the photogrammetric analyst with step-by-step procedures, without detailing how the procedures have been derived.

In this manual, we have used specific aspects of single-photograph perspective to present a fresh review of techniques commonly used in photogrammetric analysis of one-, two-, and three-point perspective photography. Normally, these techniques are elaborated upon in books and journals on architectural drawing, and are used in transforming orthographic views into perspective views (McCartney, 1963; and Walters and Bramham, 1970). We have taken these techniques (with a few modifications) and have applied them to transforming close-range perspective photography into orthographic views (Williamson, 1986, Brill and Williamson, 1987; Williamson and Brill, 1989). The close-range perspective photograph, typical of forensic photography, is actually an auxiliary perspective of the area or object of concern. Given the auxiliary view, the photogrammetric analyst can use perspective techniques to construct the orthographic plan and elevation views (with all due concern for accuracy) - Fry, 1969; Kelley, 1978-1983; Williamson and Brill, 1987-1989.

The use of single-image perspective techniques is not widely discussed in photogrammetric books or journals (Slama, 1980; Wolf, 1974; Moffitt and Mikhail, 1980). We hope the present work will make the procedures easier for the photogrammetric analyst using single images (particularly to forensic close-range applications), and also that it will stimulate further discussion of the application of perspective techniques in close-range photogrammetry.

Special thanks are given to the people who supported the authors in the preparation of this manual. Specifically, Jennifer L. Williamson for typing, proof reading, and keeping the coauthors working together; the coauthors (if we may pat ourselves on the back for working and reworking the many solutions), and nameless members of American Society of Photogrammetry and Remote Sensing for their astute comments on the text and the graphical procedures; and the graphical illustrators and format/style editor — Elaine and Gary Roland and Deborah Conn — who put up with our particular type of geometry and the many changes to the drawings and text. And to the many people who over many years, were willing to talk about single-photo photogrammetry as if it were not some dinosaur or white elephant. Notably among the latter are Gordon G. Gracie (New Brunswick, Canada), Cyril P. Kelley (Ontario, Canada), and Carl G. Mann (Maryland, USA).

DIMENSIONAL ANALYSIS THROUGH PERSPECTIVE

Chapter One
A Beginning

1.1   Introduction

This manual is intended for the professional photogrammetrist who desires a summary of methods of extracting three-dimensional measurements from single photographic images, whether these images were planned or unplanned, and irrespective of whether the camera operator was skilled or unskilled. Single-frame optical and video photographic imagery of man-made objects can be used in conjunction with perspective geometry to provide accurate dimensional analysis of an object of interest. Because each image or photograph is unique, not every procedure included will be suitable for all imagery (Kelley, 1978-1983). One image may have "strong" geometry for a certain method, but another image may have "weak" geometry for that method, rendering the method unreliable. It is hoped that enough of these idiosyncrasies of strong and weak geometry solutions have been noted in this manual to make the user cautious in selecting an approach for dimensional analysis. The following sections in this chapter will provide discussion on background and the setup of this manual.

1.2   Background

Normally we do not expect the average automobile driver to know how to work on the engine of the automobile, but we do expect it of the automobile mechanic — he is the expert. The same situation applies to camera users (photographers) and photogrammetrists. We do not really expect the average camera user to know mechanics and optics in order to use a camera. We do expect the average camera user to know the brand name of the camera, how to put film in it, and the fundamentals for operating it. On the other hand, if the operator is using the camera for a specific task such as dimensional analysis (photogrammetry), the operator's knowledge should extend beyond these fundamentals.

"Murphy's" basic rule of close-range dimensional analysis seems to be "never allow the photogrammetrist near the camera, nor is he to be present when collections are made." This means that the photogrammetrist must communicate his needs to the camera operator, and the most likely opportunity would be through field trials. These field trials should produce imagery with which the photogrammetrist is able to demonstrate to the camera operator the photogrammetric purpose of the collection. The photogrammetrist will show the camera operator how the collected imagery will be used and what collection parametric values are required.

The camera operator should keep a log of camera and frame imagery collected. The camera log should contain, if possible, the information listed in Table 1-1. Some of the log items can be provided before the operator travels to the collection area — and should be logged in advance. Other items are noted at the time of collection and still others may be filled in after the fact. As lengthy as this list appears, more items could be added for special tasks.

The major problem with a collection log is keeping it current. The only legitimate reason why the log information is not completed shortly after collection is that the film has been damaged beyond usefulness. In this case, the appropriate entry is made in the log and the collection is made again, if possible. Oh, by the way (a famous photogrammetric management expression), to avoid loss of permanent records and also transcription errors, the permanent log should never be taken into the field, and field notes should be filed along with the official recorded log.

When the film is removed from the camera, it should be marked with the page number from the log to make the log information accessible with that collection. It is often appropriate for the operator to write the information on a piece of paper (or on a form) and to photograph it as the first frame of that collection roll. A sample form for this mini-log appears in the Appendix. This expedient places the information where it is inseparable from the original negative. Even a film record of the roll number, camera and lens information would be useful in dimensional analysis, if the items marked with an asterisk (Table 1-1) are not available. This procedure is applicable to hand-held close-range collections, although it may be possible with other close-range small-format imagery.

As important as these precepts may be, camera operators do not often keep careful records. Even in the most carefully planned collection, frame imagery is collected with little or no information about the camera position or its relative orientation to the object imaged. One reason for this lack of data is that photogrammetrists are viewed as "wizards" able to supply the missing information. Another reason is that, although the operator knows he is supposed to record the information, the procedure was not

---

Table 1-1. List of Items for Camera Log

Camera name and/or manufacturer's name
Camera serial number
Lens serial number
Date(s) of collection *
Time(s) of collection (this time should be corrected to the Naval Observatory time standard)
Nominal focal length of lens used *
Range (distance) from camera to object
Any known dimension on or near the object of interest *
Calibration report number and/or calibration report*
Number of exposures and rolls of film
Number of camera stations
Latitude of object
Longitude of object
Weather conditions
Availability for future collections
Relative position of object to each camera station
Height of camera operator (Camera height) *
Was tripod used? Can it be used for future collections?
Complete physical description of object, including colors
Purpose of object
Was object (or parts of object) static / dynamic? *

* essential items

---

ingrained as habit during his/her training. Hence the procedure is forgotten as events unfold. If correct record-keeping is not made an integral part of the training, we cannot expect it to be practiced consistently.

From the point of view of the photogrammetrist, the lack of data poses a challenge to frame-imagery analysis. It is not the only challenge. Often, imagery received for analysis has been cropped or even derived from some publication or press release. Printed imagery is particularly difficult to use in accurate dimensional analysis. The halftone or screening processes of newspapers and magazines tends to distort the geometry of the original imagery, sometimes enough to make it impossible to recover the geometry required for dimensional analysis. Each generation of reproduction of the original imagery introduces new distortions and degradations. The effect of these distortions on the analysis depends on the geometry of the original imagery and the geometry of the imaged objects. These dependencies render each image within a collection unique, even though taken with the same camera and apparently from the same camera station.

Since the necessary information is not always available, the photogrammetrist must learn to rely more on the geometry in the imagery and less on accompanying data. Using the perspective geometry of single-frame imagery is not a new approach; it is the only approach. Using perspective geometry is possible only if the user is aware of the man-made geometry captured in frame imagery — particularly straight lines (which are not usually found in nature). The useful linear features (properties) of man-made objects include parallel lines, orthogonal lines, and connecting planes. By using these geometric properties the photogrammetrist can determine dimensions of the object of interest. The mathematical tools needed to determine the dimensions include geometry (plane, solid, descriptive, and analytical), algebra, trigonometry, matrix algebra, and vector analysis. We are talking about using these mathematical tools. If you know these things, or want to learn them, that is great. In this manual we use them as defined tools, and we will probably do it without your being aware of the specifics. This manual is intended to be a working tool for the photogrammetrist, who must obtain dimensions from imagery and needs to do it now!

1.3 Manual Setup and Description

Dimensional Analysis Through Perspective is meant to be used as a basic reference (working tool) from which a photogrammetrist can complete the dimensional analysis of single- and multiple-image perspective imagery. The single-photo perspective analysis is divided into graphical and analytical procedures, and either procedure or a combination of both may be used. To facilitate the use of analytical dimensional-analysis methods, we present programs for these methods suitable for desk-top computers. Although this manual contains a step-by-step procedure for the single-photo analytical analysis, the description is not encumbered by detailed mathematical descriptions. Normally, manuals such as this one go into great detail on the analytical procedures, assuming that the reader has already completed some sort of dimensional analysis of imagery and has measured data for that analysis. However, the most difficult and time-consuming work is identifying the geometry and the image points to use, measuring the data, and double-checking this work to make sure all the required data has been obtained. The time spent in doing this work is very important, critically affecting the accuracy of the results. If solutions are obtained without concentrated effort on identification and mensuration, they are likely to contain errors that waste the photogrammetrist time. The authors feel that just to solve a problem requires enough labor without having to develop the procedure. We hope this manual will provide ready references to methods, allowing time to complete a task accurately without having to develop the procedures.

Attention must also be given to the significance of numbers. In using this manual you may first appreciate the importance of numerical significance when you are trying to follow a step by step reference. We suggest that all analytical values be recorded at least five places to the right of the decimal, with the exception of trigonometric values which we suggest be recorded nine places to the right of the decimal. Internal calculations of various computing instruments will vary, as will their round-off and truncation formats which will cause solutions to differ. We use this recorded information format only to clarify the calculations that follow the recording of data. A computed number can be no more significant than the input values from which it is computed, and probably one decimal place less significant. The significance of a number is its implied accuracy. In photogrammetry, a number without an accuracy statement is worthless. So, in suggesting these methods of recording data, we have violated some important rules for the significance of numbers; however, the final answers should be presented correctly and as accurately as the value(s) used for scale.

This manual is unusual in containing both analytical and graphical procedures. It is separated into six sections. The first section, which includes the Table of Contents, Lists of Figures and Tables, and the Preface, is designed to help the reader locate reference information. The second section, Chapters One and Two, presents information concerning this manual, and conventions and classification of single-photo photogrammetry. The third section, Chapters Three through Seven, contains the single-photo graphical procedures. Chapter Eight is a listing of definitions and serves as a separator between the graphical and analytical procedures. The fourth section, Chapters Nine through Twenty, contains references to single-photo analytical procedures. The fifth section, Chapters Twenty-One and Twenty-Two, contains a synergy of graphical and analytical application methods for dominant and combined perspective geometries, and the conclusion to this manual. Section six is the Appendix with additional reference information and 32 example problems.

Unique solutions have been included as needed. The graphical solutions in Chapters Three through Seven have a limited number of graphical presentations for both the text description and the examples. Therefore, the labels in these graphical presentations are extremely important, as one graphical perspective illustration will contain many steps of the solution procedure. Each single-photo analytical chapter begins with a description of the procedures to be used, and has step-by-step procedures listed for your immediate reference. The analytical examples are described with just enough detail to allow you to follow the methodology. By reworking each example you can learn the procedures applied.

DIMENSIONAL ANALYSIS THROUGH PERSPECTIVE

Chapter Two
Conventions and Classifications

In this chapter we will consider the general requirements for some of the geometric conventions of perspective analysis. In particular, we will discuss the image and object space coordinate conventions, and the rotation of the image relative to the object space coordinate system. Finally, we will discuss some very important history concerning projections.

2.1   Axis Convention

The relationship between the image and the object-space coordinate system is shown in Figure 2-1. Since the problems treated in this manual involve relative positions and dimensions rather than absolute positions, any convenient object-space coordinate system (right-hand system) can be chosen. The object-space coordinate system shown in Figure 2-1 is the system chosen for all procedures and examples in this manual. In this coordinate system, the positive Z-axis is directed vertically upward, the positive X-axis is directed toward the right and away from the camera station, and the positive Y-axis is directed toward the left and away from the camera station. The system is orthogonal, so each axis is at an angle of 90° from the other two axes.

The object-space coordinate system (upper-case letters) is positioned and oriented in such a way that the camera station coordinates, Xc and Yc, are always negative. This assumes that the origin of the coordinate system (0,0,0) is on the object of concern. The camera station coordinate Zc is either positive or negative, depending on whether the camera station is above or below the XY coordinate plane.

Objects, visual rays, and the camera position of primary concern during the taking of a photograph are conveniently described in a particular image-space coordinate system. The image-space coordinates (lower-case letters) form a three-dimensional cartesian system with an origin at the lens nodal position. In this coordinate system, the image plane itself is a plane of constant z (where $z = \mathbf{f'}$, the effective focal length at the time of the exposure), and spanned by the orthogonal coordinates x and y. The x-axis is chosen by convention to be parallel to the bottom edge of the frame (i.e., the edge closest to the ground). The camera station (synonymous with the lens nodal position) is a distance **f'**, along the positive z-axis, from the image plane. The z-axis of the image coordinate system extends from the origin (principal point - **pp**) to the camera station, which is the effective lens nodal position.

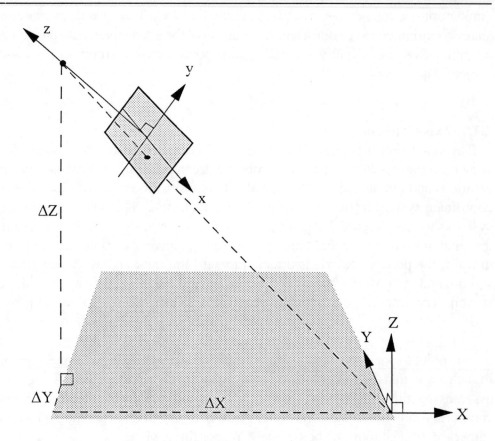

Figure 2-1. Image and Object Space Axis Relationship

Photogrammetrists do not have immediate knowledge of the absolute image-space coordinates to points in an image-space, even after the photograph has been taken, and so measurements begin in an arbitrary two-dimensional mensuration coordinate system. The origin for the mensuration coordinates need not be the **pp**, but may be any image-plane position convenient for the photogrammetrist. The position usually selected is below the bottom edge of the frame and to the left of the left edge of the frame. By placing the mensuration origin in this location and keeping the mensuration x-axis approximately parallel to the bottom edge of the frame, all of the mensuration coordinates will have a positive sign. This is helpful when using a pocket/desk calculator because the sign of a value does not become a factor in the calculations. If possible, the photogrammetrist should complete a transformation of the mensuration coordinates so that the mensuration x-axis is parallel to the bottom edge of the frame. Equations for this transformation are given in Chapter Twenty-one, along with a reference step-by-step procedure.

2.2 Rotation Convention

The graphical solutions use the rotation angles of azimuth, tilt, and swing to describe the rotation of object-space to image-space coordinates. Multiple-frame image

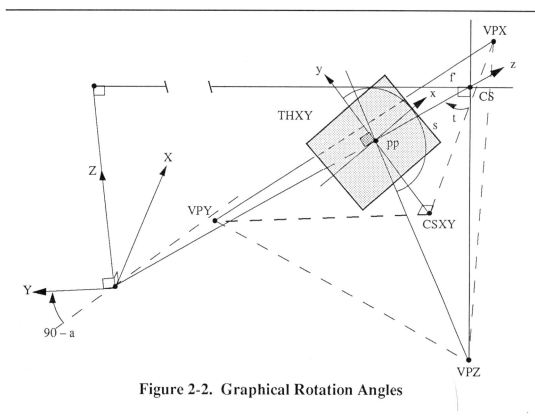

Figure 2-2. Graphical Rotation Angles

coordinate systems also exist, and use the rotation angles of omega, phi and kappa. There is a significant difference in the definitions for these two sets of rotation angles. Figure 2-2 illustrates the graphical rotations. Figure 2-3 illustrates the rotations for the multiple frame solution.

In single-frame graphical analysis, the azimuth is a clockwise rotation about the object-space Z axis, and is the angle between the object-space Y-axis and the principal plane. (Principal plane is defined by the three image points of **pp**, nadir, and the lens

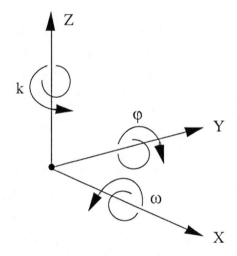

Figure 2-3. Analytical Rotation Angles

nodal point.) The azimuth angle will vary in magnitude from +0° to +90°. The tilt angle is a rotation about a line parallel to the true horizon line, and is the angle between the principal ray and the line from the principal point to the Z vanishing point (nadir point). The tilt angle will vary in magnitude from +0° to +180°. If the tilt is greater than +90°, the camera is pointing upward. The swing angle is a rotation about the image z-axis (**f'**) and is the angle between the positive image y-axis and the trace of the projection of the principal plane below the image x-axis. This ensures that the swing angle of frame imagery will not be less than +90° nor greater than +270°. If the image x-axis is parallel to the horizon line between the vanishing points X and Y, the swing angle will be +180°.

2.3 Classifications of Projections

In order to understand the geometry represented in the two-dimensional plane of frame imagery, it is necessary to have an understanding of the classifications of projections. Projections that concern the photogrammetrist appear in drawings or graphical projections. There are four classifications (forms) of projections: orthographic, oblique, axonometric, and perspective. Of primary interest is the perspective projection, but all four forms need to be understood in order to work with any one of them.

The form of projection from a planar object to an image depends on the juxtaposition of the image plane, lens, and object plane of interest. The human eye is a good example of these factors. Light reflected from the object plane is projected over some range, through the lens of the eye, and onto the image plane of the eye (the cones and rods). The same geometry is present in a camera system, with some major differences. Once the eye is closed or is moved, the projections change and are not likely to be duplicated. Light from an object plane, at some range, passes through the camera lens and strikes the film (image plane) causing a chemical reaction. Because of this film reaction it is possible to record a two-dimensional image of the three dimensional projection. It is the geometry between the three dimensional scene and the two-dimensional projection that must be understood.

2.3.1 Orthographic Projection

The first projection we discuss is orthographic. The word orthographic comes from the Greek words "ortho," meaning correct, and "graphic," meaning writing. Thus, orthographic is correct writing and to translate this further it means to illustrate something as it really is. To obtain an orthographic projection the four factors (object, range, lens, and image) must be arranged so that the projection of the rays from the object plane, at some range, through the lens, onto the image plane provides a true representation of the object. This condition is met if the image plane is parallel to an object plane, and the range is great enough to cause the reflected light rays to be essentially parallel. Engineers and architects use this type of projection all the time; it is called a blue-print. A set of plan and elevations views, are orthographic projections. It is possible to measure directly from the orthographic projection and, by applying a given scale, determine the actual size of the object.

The immediate problem is how the four factors (object, range, lens, and image) are arranged to produce this orthographic projection on the film. Unless you have studied projections, more background information is needed. People see things because of light reflection. In a conference room a large box standing in the middle of the room is seen by a group of observers, but not every observer receives the same image. The image received depends upon the position of the observer's eye (or camera) in the conference room, and the subtended visual angles of the reflected light. Observers positioned to the right see reflections from the front and left side of the box (box left is observer's right), the observers to the left see reflections from the front and the right side of the box, while those in the middle see the front of the box. Keep the observers in the same relative positions and place them and the box in a large meeting hall. Standing at the other end of the hall, the observers will see as much light reflected from the sides of the box as before, but the subtended visual angles will be smaller because the box is further away. Those standing to the right will see more reflected light from the left side than those standing to the left, etc. Now change the scene, and place the large box at the end of a football field (closer to infinity). Keep the observers in the same relative positions standing at the other end of the field. All of them will now have about the same view of the box. The same effect can be observed by keeping the box in place and moving slowly from one side of the box towards the center.

Of the four factors (object, range, lens, and image), only one has changed in each instance. The range (the distance from the object to the lens) increased. By definition, in perspective, parallel lines meet at infinity. Light rays emanating from infinity are thus defined to be parallel. In fact the reflected light does not have to be at infinity, but just a very large distance away from the lens. (Recall how straight railroad tracks appear to converge in the far distance.) This is approximated by stating that if the ratio of the range (object plane to lens) over the distance from the lens to the image plane (focal length) is extremely large, then the object plane may be considered to be at infinity. The ratio of range to focal length, together with the visibility of the object, are the deciding factors. For all practical purposes, imagery of an object with a large ratio of range to focal length will be an orthographic projection. The orthographic projection will not be true size and only the view imaged can be represented. The lens collects the reflected light and refracts the rays so that the proportionality of the object remains as a change in scale. A scale factor (or conversion factor) is required to determine the correct size of the imaged object. One way of determining the scale/conversion factor is to use the range divided by the focal length. This conversion factor will yield the true dimensions of an object at that range in an orthographic image, within the conventions of orthographic projection. To determine the other planes (standard sides, as viewed in a blueprint), the object would have to be rotated 90° or the camera would have to move 90° around the object.

2.3.2 Oblique Projection

Oblique projection involves the same four factors (object, range, lens, and image) as orthographic projection. The difference is that although the plane of the object and the image plane are still parallel to one another the viewpoint of the observer has been moved. Using the example of a box, as before, place the observer above and to the left of the box. The observer would see the front plane of the box in orthographic view, but the top and right side of the box, now also in view, would be oblique projections. The back edges of the two planes (top and side) representing the edges of the back plane of the box would still be in orthographic projection. The edges of the box where the top and side planes meet would not be true length lines, but would be foreshortened. In effect all vertical and horizontal lines in an oblique projection, parallel to the image plane, are true length lines. All other lines in a oblique projection are foreshortened; their projected lengths are related to their true sizes by scale factors that depend on the angle at which each oblique view is seen.

2.3.3 Axonometric Projections

The family of axonometric projections consists of Isometric, Dimetric and Trimetric projections. To describe axonometric projections we shall use the image of a cubical box and draw lines on it. Consider looking at the box from a position where you can see the front, top and right sides (see Figure 2-4). You, the observer, are now looking at one corner of the box and also at three other edges. Mark the center of each edge and draw a line between each *pair of* marks. There is now a line on the front side from the middle of the right side edge to the middle of the top side edge. The same applies to the other two sides (top and right). These lines describe a triangle with edges on three sides of the box. This triangle is the object plane, parallel to the image plane and to the axonometric projection plane. The lines from the corner of the box to the corners of the triangle define the axis lines. When the three sides are equal, as just described, the axonometric plane is isometric and the scale along axes of the axonometric plane are the same. Tilt the axonometric plane so that one of the axes becomes longer or shorter than the other two (which remain equal) and you have the dimetric axonometric projection. The scale along the two equal axes is the same. Tilt the projection plane again so that none of the three axes have the same length or scale and you have the trimetric axonometric projection. Remember in all three instances of the axonometric projections you still have the same four factors: object, range, lens, and image.

Three of the four forms of projection have been presented in limited form. The theory of each form is worth remembering, as each has a similar form within the perspective forms of projection.

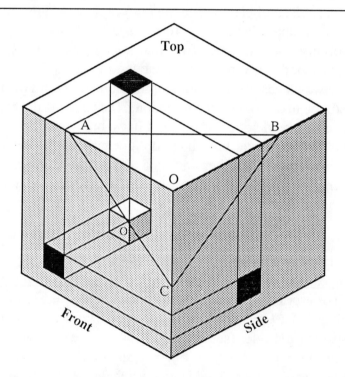

Figure 2-4: Cubical Box as Axonometric Projection

2.3.4 Perspective Projection

The significant difference in perspective projections, when compared to the other three forms of projection, is that the object is no longer considered at infinity. The effect is that the value of the ratio, range to focal length, will no longer be considered large.

All light rays from the object will no longer be parallel, but will converge at the lens. Only one light ray will be perpendicular to the image plane and that is the one coincidental to the principal ray of the lens. This last fact is very important in solving of perspective projections.

At this point in the discussion of projections the relationship to photogrammetry is to be considered. The casual observer may become confused when references are made to exterior parameters (object space) and interior parameters (image space). It should be noted that the lens is considered a common point in both sets of parameters. The lens may be considered a pivot point about which the camera rotates. The camera station, station point, and lens all refer to the same point in object space. Yet, lens and nodal points may refer to the same image position. Earlier, in the discussion of coordinates, the principal point was referred to as the image plane origin with coordinates 0,0,0. The origin of the image system is along the principal ray at a distance from the last nodal point equal to the effective focal length of the lens at the instant of exposure. The origin is the position of the principal point in the image plane.

2.3.5   One-Point, Two-Point, and Three-Point Perspective

The difference between perspective and nonperspective images is seen in optical frame imagery of identical objects placed at increasing ranges from the camera station. The projection of the farthest object onto the image plane is much smaller than the projection of the closest object. In the juxtaposition of object plane, lens, image plane and range, infinite combinations of projections are possible. There are three kinds of perspective projections to consider: one-point, two-point and three-point perspective. All three projections can be present in an image, depending on the position of planes and lines relative to the image plane.

2.3.5.1 Scale

In the forms of projections discussed previously, the distances from:

a. the image plane to the lens at infinity, and

b. the object plane to the lens at infinity

have been of no consequence. The reason for this is that all the visual rays have been parallel, and thus the image size has been unaffected. However, when the range is well within infinity, objects of different size can appear the same size on the same image plane, because the range of each object must be considered independently. It can also be shown that objects of the same size placed in different positions will be of different sizes on the same image plane. The effect is caused by the change in scale of the objects which are now in different object planes. Remember, the scale factor is the ratio of the range to the focal length, and as each object plane has a significantly different range and therefore a different ratio (scale factor), the image size changes accordingly. This again leads us to the three new perspective projections: one-point, two-point, and three-point perspective.

2.3.5.2 Characteristics of Perspective

The characteristics of each of the three perspective projections are best explained by graphical methods. Not only is it easier to show the perspective points graphically, but possible errors are easier to understand. Historically, perspective was not represented in art until about 1400 AD. Earlier paintings show no appreciation of perspective, as parallel lines do did not converge. In the previous discussions of parallel light rays converging at a point, it was stated that the point was an infinite distance away. It was also stated that for us to appreciate this geometry of perspective projections, parallel lines will always converge. The point of convergence is dependent upon two situations. The first is the separation of the parallel lines and the second is the angle between the parallel planes containing the parallel lines and the image plane. If the parallel lines are not close together, it is likely they will converge close to the imaged object. If the angle between the two planes (image and object) is small, the convergence of the parallel lines will probably be far away. Needless to say, there are infinite combinations of these two situations.

2.3.5.3 One-Point Perspective

In the presentation of one-point perspective the two situations described above seem unique. To describe one-point perspective, imagine that you are standing in front of a brick apartment building, and the distance between you and the building is less than the width of the building. You are standing on the center line of the width of the building. You can see only the face of the building. You take a picture of the exterior of the building by pointing the camera directly along the center line of the building so that the image plane and the front plane of the building are parallel. After your film is developed and a photograph made, you would be able to confirm that:

a. the verticals of the building remain vertical in the photograph;

b. the horizontals of the building remain horizontal; and

c. consequently right angles on the building remain right angles and all shapes on the face of the building remain the same on the photograph.

These findings are identical to that of an orthographic projection, although the station point in perspective is a finite distance from the object. (For an orthographic projection the station point is at infinity.) Well, seeing is believing, so the photograph must be an orthographic projection! Wrong, the photograph is a perspective projection that has the same properties as an orthographic projection. The ratio of the true image (orthographic) is proportionally the same in this plane of a one-point perspective projection.

It is this principle that is used in vertical aerial photography, where the image plane of the aircraft camera is parallel to the ground plane. Stated another way, the aircraft has no tilt angles when the photograph is taken. Therefore, all equal distances on the ground will be equal on the image or, in other words, the same scale will apply over the entire area of the photograph.

2.3.5.4 Two-Point Perspective

The presentation of two-point perspective will be discussed in more detail in Chapter Six, but we introduce pertinent definitions here. The major defintions to remember about two-point perspective imagery are:

a. The tilt angle, by definition, is 90°.

b. Given known diagonal angles, the camera station, principal point, and effective focal length can be determined.

   c. Lines perpendicular to the horizontal plane are truly parallel lines in the image, and are perpendicular to the XY horizon line.

   d. Plan and elevation views can all be drawn relative to and above the XY horizon line.

2.3.5.5 Three-Point Perspective

Three-point perspective is unique in that no line in the imagery that describes the orthogonal geometry is parallel to the image plane. To have three-point perspective in an image is to have all the tools necessary, except scale, to perform the photogrammetric analysis. Three-point perspective will be discussed in more detail in Chapter Seven. The major defintions to remember about three-point perspective imagery are:

   a. The azimuth, swing, and tilt angles are defined by the geometry of the object and the position of the camera.

   b. All of the parametric values required to complete the three-point analyis can be determined in a well defined step-by-step analysis once the three vanishing points are determined.

2.4 Conclusion

Probably the most consistent fact about single-photo imagery is that every collection is unique. So unique in fact that photogrammetrists are always using the same techniques, but packaged differently for each collection. Is the imagery one-, two-, or three-point perspective? Is the vanishing point geometry orthogonal? Are there sufficient parallel lines to locate a vanishing point? Can the end points of the known dimensions be located? Are all the building blocks available or does the wizard of photogrammetry have to call on one of his many incantations to conjure up a solution for the required parametric values? Look for the answers to these questions and many more in the following chapters.

DIMENSIONAL ANALYSIS THROUGH PERSPECTIVE

Chapter Three
Vanishing Points, Camera Station, and Principal Point

In this chapter we approach the subject of vanishing points (a sense of humor may be required). For those who know the subject, the material may seem redundant, but go along with those who are not as knowledgeable in perspective theory, and you may find this chapter is necessary. In previous chapters we introduced the concepts of projections and perspective. Here we place the emphasis on locating vanishing points because of perspective with the hope that this reflects the experience of the authors and the extreme flexibility required by photogrammetrists. The first part of this chapter will review the key elements of vanishing points and the graphic skills required. Then we will cover primary vanishing points and their location. The location of other vanishing points, such as diagonal vanishing points, will be discussed. The chapter closes with a discussion of locating the image camera station and principal point in two-point and three-point perspective analysis.

3.1 Introduction to Vanishing Points

The term "Vanishing point" is a descriptive phrase meaning a position where parallel lines seem to converge and disappear. In reality a vanishing point is the infinite distance position where all parallel lines converge. To illustrate that a vanishing point is not a black hole nor a vanished point it should be remembered that the star about which Earth orbits is the vanishing point for parallel shadow lines. And our star is neither black nor a hole and is easy to locate. A vanishing point is, by definition, a point at which imaged parallel lines converge. This looks like a contradiction in terms, "intersecting parallel lines," but it is not in the geometry of photogrammetry.

In the two-space representation of a three-space object (a photographic image), parallel lines of three-space are displayed as converging lines. The image most easily imagined is to be standing between two railroad tracks and seeing the rails converge in the distance. The perspective distorts what is seen because the railroad tracks most certainly do not converge in the distance, they only appear to converge. This appearance is due to the change in scale as the distance between the observer and the object increases. This is how we perceive distance: objects of the same size appear to get smaller the farther away they are. If we could determine the scale at the position where parallel lines seem to converge, and if we could measure the actual separation between those lines, correcting for scale, we would then find the separation of the rails to be consistent with the observer's position. The lines are truly parallel. As we approach distant objects we are aware of the change in apparent size. There is a limit to the distance at which we can perceive a change in scale and that distance is about 2200 feet. In other words, objects we can see that are further away than 2200 feet do not seem to get closer as we walk towards them. If we approach these distant objects at a very rapid rate, or if the object is very large, the rate of change in scale would be noticeable. Driving across the state of Kansas towards the state of Colorado, with the Rocky Mountains in the far distance, is just such an example. At first there is no apparent change in the size of the mountains. Later, when well into the state of Colorado, there is a very noticeable change in the size of the mountains as the range decreases. The effect is caused by binocular vision (stereo vision).

3.2   Graphics Skills for Vanishing Lines and Points

Now that a very lengthy description has been given for vanishing points, a few incidental details should be discussed. To reproduce a three-dimensional scene on a two-dimensional surface (paper) requires a certain graphic arts skill - drafting. (This, of course, assumes a single-perspective photograph). Drafting is not a natural or instinctive skill - it must be acquired through knowledge and lots of practice. Of primary importance is the use of a pencil (a scribe or an ink pen, etc.) and straight edge. Graphically we want to make true representations (orthographic projections) of three-dimensional space on paper. We want to place on paper this true representation of what we have already collected on film and printed as an enlarged paper print. The image our brain received when we saw the object, is rarely of any use because it was in real time and that image is often distorted by our ability to recall the original scene. Before we digress too far, just remember the basic principles of a camera and of how we see are the same, and we can relate what we saw to the photograph. Now, back to developing graphic skills for vanishing lines and points.

When seeing the railroad tracks in our example, we see a real set of straight lines (the tracks) converging to a vanishing point. Likewise, we can also see horizontal parallel lines of buildings converge when we walk down a street, and we extend these imaginary lines from the building to the vanishing point. The amazing thing about our imaginary line is that it is straight and of constant thickness from beginning to end. When we draw such a line, we want to duplicate that imaginary line as exactly as we can. In doing this we are going to use a photograph, a straight edge, drafting paper and pencil. The photograph provides instant recall. The straight edge is a method of defining the extended line, and the paper and pencil are materials for reproducing the extended line (French, 1958).

As stated, the straight edge is a physical reproduction of our extended imaginary line. The photograph is placed on the paper. By placing the straight edge in coincidence to the line of the photograph and using the pencil along the side of the straight edge we trace (draw) a line. At whatever attitude we hold the pencil when we begin is the precise position the pencil should be held the entire time the tracing is made. If not, the line will not be straight and most likely will not pass through the unknown location of the vanishing point. The fact that it does not pass through the vanishing point will not be known until after we have drawn several lines and they do not intersect at a common point. In fact, if our drafting is really sloppy, the lines will converge in pairs. This means that there will be a lot of two-line intersections and very few, if any, intersections of three or more lines. If our sloppiness is truly random we could draw a polygon using the loci of all of the two-point intersections, and then use the center position of the polygon as the best estimate for the vanishing point. Normally the vanishing-line drafting error is biased by the length of the image line, the length of the drafted line, and the photogrammetrist's experience in drawing lines. Only extreme care and time will reduce errors.

There is a method for reducing repeated vanishing line drafting errors, and that is to avoid drawing of pencil lines. Instead, the photogrammetrist should use long strips of prepared clear acetate. The acetate should be about two inches wide and 36 inches long, with an accurately scribed straight line on it. These lines are placed over the photographic image in place of pencil lines. When three or four of the scribed lines have been placed over the image vanishing lines they should intersect at the vanishing point. If they do not intersect precisely it may be that they were placed incorrectly (maybe even scribed incorrectly) and it is easier to adjust them than it would be to adjust a pencil line. Once the vanishing point has been located in this manner it is a simple matter to mark its position and check its accuracy using the acetate strips or a straight edge and other parallel lines in the imagery.

3.3   Primary Vanishing Points

It is unusual for nature to provide the object-space geometry necessary for perspective analysis; in general only man-made objects have this geometry, even though the form of object-space geometry is dictated by a physical law of nature - gravity. Man-made objects in object-space are built of horizontal and vertical intersecting planes. Lines through the intersection of these planes are parallel. A condition of this object-space geometry is that parallel lines will appear to converge at a common point at infinity when viewed from a perspective center. An optical lens is such a perspective center; e.g., your eye contains a perspective center. We see in three dimensions, we build in three dimensions and therefore there are three primary vanishing points and of course a projective center.

The three dimensions we will define are mutually orthogonal, meaning each dimension has a coordinate axis that is at right angles to the other two coordinate axes. This defines the geometry of our photogrammetry, but it is possible to determine this geometry from non-orthogonal planes that are perpendicular to the horizontal plane. Each orthogonal coordinate axis has a vanishing point associated with it. The vanishing points of a photograph also have a projective center. The lens (camera station) is the projective center through which all lines must pass when projected from object-space to image-space or vice versa. The projection of the principal ray (of the lens) through the projective center onto the image plane is the principal point (PP). The principal ray is perpendicular to the image plane at the **pp**. The four primary vanishing points projected on the image plane are Center (**pp**), Right (**VPX**), Left (**VPY**), and Down/Up (**VPZ**). When any three of these four points have been located, they may be used to locate the fourth image position; e.g., if **VPX, VPY, VPZ** are known, **pp** can be found; if **VPX, VPY, pp** are known, **VPZ** can be found, and so on.

3.4   Other Vanishing Points - Diagonals

Other vanishing points may be located on a horizon line or on lines perpendicular to a horizon line. The most common of these are the horizontal and vertical diagonal vanishing points. A standard horizontal diagonal is formed when the geometry in a horizontal plane is in the form of repeated rectangles. (A square is a unique rectangle.) The square tiles in a floor pattern, the expansion joint pattern in concrete roadways or airport taxiway, sidewalk patterns, and pond patterns in fish hatcheries are a few examples. All horizontal diagonals, diagonals in the XY plane, will converge to vanishing points on the XY true horizon line. The same pattern scheme applies to vertical diagonals. The best examples are the diagonals of common windows and doors. The windows of the same geometry will have a vanishing point, as will the doors of the same type of geometry. The vanishing points for these diagonals of the XZ and YZ vertical planes will be on the XZ and YZ true horizon lines. For two-point perspective, these horizon lines are perpendicular to the XY horizon line and through the respective X or Y vanishing point. Because of the unique properties of the geometry in three-point perspective, the horizon lines between the major vanishing points are used to locate the respective diagonal vanishing points.

The sun as a vanishing point (see Figure 3-1), presents a unique location. Parallel shadow lines on the XY horizontal plane will converge to a vanishing point on the XY true horizon. The inclined shadow lines will converge on a line from VPZ (for three-point, and perpendicular to the XY horizon line for two-point) through the Sun's horizontal vanishing point. This is an example of diagonals in planes not parallel to the major vanishing point planes. Care must be exercised to use the correct geometric principles with the perspective scene. The commonality between these vertical planes is that they are perpendicular to the horizontal plane, although they are not mutually orthogonal with the major coordinate axes. The procedures for finding and using these vanishing points are distinctly different between two- and three-point perspective solutions.

3.5   Camera Station and Principal Point - Two-Point Perspective

The rule of interest for horizontal vanishing points concerns the angle formed at the camera station and the location of the principal point. So, before we discuss these vanishing points further, it is important that the geometry of the camera station (projective or perspective center) be discussed. To get a clear, concise concept of the geometry involved with the image position of the camera station and how the formation of this geometry is so important, we shall start with two-point perspective for which the following conditions are true:

1.   There are only two major vanishing points, the vanishing point right (**VPX**) and the vanishing point left (**VPY**). The third major vanishing point is at infinity either up or down.

2. The principal point is located on the true horizon line XY.

3. There are still three true horizon lines in two-point perspective. The true horizon line XY extends through the major vanishing points of VPX and VPY. The XZ true horizon line is perpendicular to the true horizon line XY and passes through VPX. The true horizon line YZ is also perpendicular to the true horizon line XY and passes through VPY.

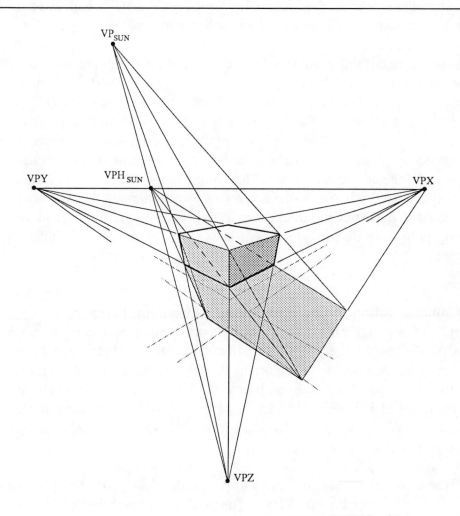

Figure 3-1: Sun's Vanishing Points

4. The camera station is located on the circle having as its diameter the distance between VPX and VPY, and it is located on the line perpendicular to the XY true horizon through (or from) the principal point. For convenience the camera station is located on the circle intersection point below the XY true horizon line (in true two-point it could just as well be above the horizon line).

In this example the vanishing points, left and right, were found using the parallel lines of a checkerboard as shown in Figure 3-2. In order to demonstrate what is happening the perspective lines will be drawn and then the explanation provided. Remember, according to condition Number 2 above, the principal point is on the XY horizon line.

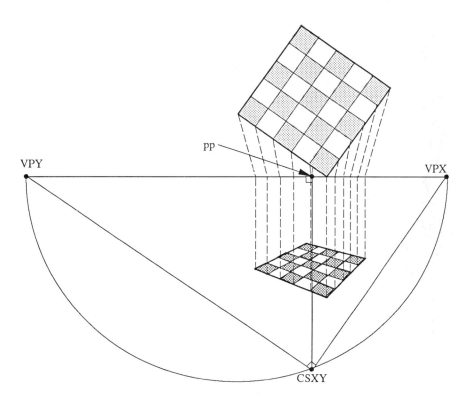

Figure 3-2. Two-Point Perspective - Checkerboard

This means the camera is level (tilt angle is 90°) and the vanishing points, along with the camera station, lie in the horizontal plane. Since a checkerboard is made up of squares, the two sets of parallel lines defining the vanishing points are orthogonal (at right angles to each other). Therefore, the angle between the lines from the camera station to the vanishing point is a right angle. This is an important concept to remember. The angle between the lines to the vanishing points from the camera station will always equal the angle between the lines used to locate those vanishing points. Because of this geometric condition, a true plan (orthographic) view of the checkerboard can be drawn between the lines to the vanishing points at the camera station. To keep from obscuring the image with a line drawing, the orthographic view is generated on the horizon line between the vanishing points (see Figure 3-3). The view is orthographic, so all lines and all angles will be true to scale. Look at the checkerboard as it is drawn true view. The square formed at the camera station has a diagonal at an angle of 45°. When this diagonal is extended, it will intersect the XY horizon line at the vanishing point for the horizontal diagonals of the checkerboard. There is only one point on the circle — the camera station — from which that geometric relationship will be true.

"So what!?" you say, "It is easy to find the camera station as it is stated in condition Number 4; you just intersect the circle with a perpendicular line to the XY horizon line from the principal point." Without a full format image in which the center of format is the best estimate of the principal point, we would like to see how you located that principal point. Condition Number 2 just stated that the principal point is on the horizon line, not how it was located. Now we have a way to locate the principal point and the camera station in two-point perspective. We use a diagonal, of known rotation angle, in the horizontal plane. The way to do this graphically is to locate the diagonal vanishing point on the true horizon line by extending the diagonals of the squares. Then on a separate piece of clear acetate scribe the vanishing point angle formed at the camera station (90°) and also scribe the line of known rotation angle. In our checkerboard example this angle would be 45°. The intersection of these three lines is the camera station.

To locate this camera position on the XY perspective circle one condition must be satisfied: the respective lines must pass through the respective vanishing points and the intersection point must lie on the circle. When this condition has been met the position of the camera station will be located. In addition a perpendicular to the XY horizon line through the camera station will locate the principal point on the XY horizon line. From just that one set of diagonals, with known rotation angle, we were able to locate the camera station and the principal point. Actually this two-point procedure is not easy, and is explained in detail in Chapter Six.

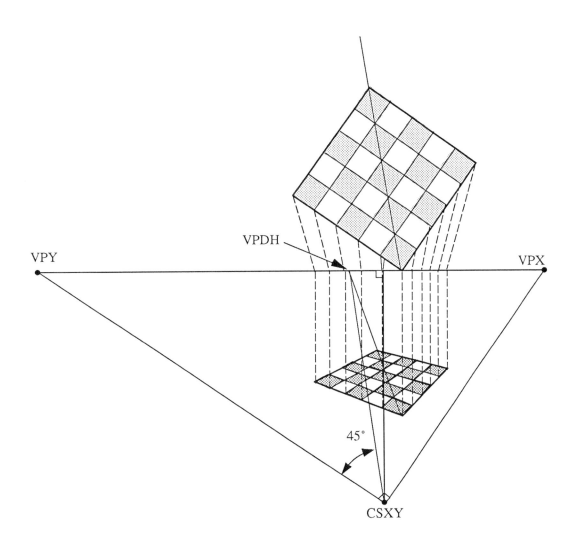

Figure 3-3. Two-Point Perspective - True Representation

3.6   Conclusion

The graphical determination of vanishing points in done by intersecting perspective parallel lines, and does require a specific skill in drafting. The perspective parallel lines can be formed by building lines, windows, doors, or the diagonals formed by the same constructions. Shadow lines, horizontal and diagonal, can also be used to locate diagonal vanishing point of the sun. Two point perspective imagery with horziontal or vertical known angles without full-format can be used to locate the prespective **CS** and **PP**. All of these methods have been discussed and are found useful in the examples described in the appendix.

DIMENSIONAL ANALYSIS THROUGH PERSPECTIVE

Chapter Four
Focal Length and Rotation Angles

In this chapter we will cover effective focal length and rotation angles. The first section covers the major distinctions between nominal and effective focal lengths, what they are, and the best ways to determine the required focal length for the imagery. The final section describes the three rotation angles: Azimuth, Tilt, and Swing, and how they can be determined.

4.1   Focal Length Values

According to the American Society of Photogrammetry Manual of Photogrammetry, Fourth Edition, the focal length is the "distance measured along the optical axis from the rear nodal point of the lens to the plane of critical focus of a very distant object." The definition continues with a description of nominal focal length as "an approximate value of the focal length, rounded off to some standard figure, used for the classification of lenses, mirrors, or cameras." These two definitions are of primary concern to the photogrammetric analyst because the first (in effect) describes the calibrated focal length, or the infinity focal length, and the second describes the focal length usually supplied with the camera and therefore with the imagery. Neither of these is the effective focal length (f') value required for close-range single-photograph perspective analysis. The close-range f' is the effective focal length at the time of the collection, and is defined as the distance from lens rear nodal point to the image plane. For all practical purposes this definition is the same definition as described above for a very distant object.

Oh, by the way, you must remember that when working with an enlargement (which you always do when working graphical solutions) the f' required is the f' as it relates to the enlargement. Instruments used to measure image coordinates for analytical solutions are usually measuring a contact postive, and the enlargement value is unity. All references to f' in this manual relate to the effective focal length value required for the solution under discussion, and there is most likely no reference to an enlargement value.

|**Note:** An assumption applied in this description, must be clarified. In close-range photogrammetry the typical camera is assumed to be a camera with an adjustable focus. Yet, it is quite possible that the photographic collection was made using a camera with a fixed lens. In that case the manufacturer's nominal focal length, and the collection effective focal length are the same. Also, it was traditional that cameras used in photogrammetric collections were calibrated metric cameras. Modern times have changed this situation, and more often than not the camera is some off-the-shelf non-calibrated camera. For the purposes of this manual we are assuming the camera has an adjustable lens for focusing on the object of concern. |

Let's digress for a minute and discuss some aspects of close-range perspective, such as the effect on perspective through focusing the camera. A quick analogy is that of an observer. For an observer to be able to discern separation of depth detail in a number of objects, the objects should be no farther away from him than about 2200 feet (671 meters). These numbers will vary according to the observer, but this is our example. (We assume the size of the objects is large enough for them to be seen and recognized.) By separation of depth detail we mean which of two identical objects is closer to the observer and which is farther away. The separation between objects will be difficult without prior knowledge of their relative sizes, and at large distances it will be difficult to see the depth separation between any size objects. This is analogous to what is happening with the camera. For close-range focal lengths (about 23mm to 135mm) there is perspective, which relates to a change in scale according to a change in range (depth). As the range to the object changes, without changing the focal length, the less perceptive is the perspective; e.g., the smaller the change in scale with change in depth the more the image appears to have parallel lines, until the objects seem to have no separation in depth and appear as an orthographic projection.

Some camera lenses, like our eyes, are adjustable in order to focus on the object at whatever distance the object is away from the lens/camera. As the lens is focused there is a definite change in the position of the lens, and therefore a change in the effective focal length (see Chapter Three of the Manual of Photogrammetry, Fourth Ed.). To provide yourself with an estimate of the effective focal length, the rule of thumb is: for long-range collections the nominal focal length can be used as the **f'**; for close-range collections the nominal focal length has increased by about 2 percent to become the **f'**. All calibration values of focal length are usually for extreme long-range collections (infinity), and the nominal value of the lens is usually a very close approximation to the calibrated value. All close-range perspective imagery will be acquired with an effective focal length that is longer than the manufacturer's nominal focal length. The change in the nominal focal length can amount to an increase of about 2 percent. A *change in a nominal focal length of 50 mm could introduce a 2 to 4 percent error in the photogrammetric analysis.

And, what does ~~has~~ this discussion of range and focus have to do with our presentation of effective focal length? Primarily to make you aware that unless the camera is calibrated for the range of the collection, the **f'** will have to be determined by the perspective geometry in the photograph. It is possible to calibrate a camera, and develop a set of graphs from which the probable **f'** can be selected for various ranges. To discuss the methods and physics of calibration any further is beyond the scope of this manual. The point of this digression was to provide some additional information about the parametric value called focal length, which is important to the photogrammetric analyst when attempting any error analysis. The <u>Manual of Photogrammetry</u> will provide you with more detailed information and references on the subject of the calibration of cameras.

Now, back to the graphical presentation of effective focal length in perspective analysis. The solid geometry of a camera is that of a right circular cone with the effective focal length being the height of the cone (see Figure 4-1). Only a portion of the light passing through the lens reaches the film. A plate (or platen) with an opening in it is placed over the film and the light reaching the film is in the shape (pattern) of the opening of that plate. The effective focal length varies because of the adjustment of the position of the lens rear nodal point to focus the light rays on the film. For non-topographic (close-range) imagery this effective focal length can change from image frame to image frame as the distance to the imaged object changes and the focus is adjusted accordingly. This also means the scale changes, ergo the perspective changes (no significant change in scale with a change in range and the perspective has flattened out to become orthographic). This is a major reason each image in a close-range collection is to be considered unique.

Although a camera produces a two-dimensional image of three-dimensional space, a camera is a three-dimensional object. The interior parameters of a camera are three-dimensional and are fixed (constant) for each image collected. We could model the camera in three dimensions but that would not help provide a graphical solution on our

* 1-2 mm

paper. Therefore, what we do is to use the intersection of vertical and horizontal planes as hinges and rotate one into the plane of the other. For example (see Figure 4-2), in order to describe the effective focal length in the image plane (graphically) we find the intersection of the effective focal length plane and the image plane and rotate one plane onto the other. It is just like closing a door to make it part of the plane of the wall. The problem is in finding the plane of the effective focal length. By definition the effective focal length (a line) is in the principal plane. For now it suffices to say that

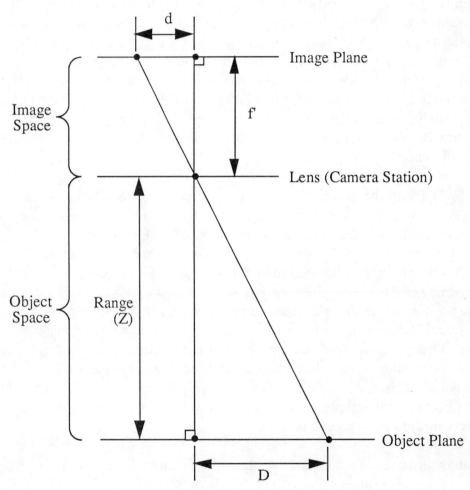

Figure 4-1. Basic Camera Geometry

the principal plane intersection with the image plane traces a line (a principal line in the image plane (called revolving or rotating one plane onto another) we should again examine the position of the effective focal length. Remember the effective focal length image plane) that is perpendicular to the true horizon, through the principal point (**pp**), and continues on to a vanishing point. The easiest method of illustrating this is to use the principal plane perpendicular to the XY horizon line, assuming Swing is 180°. This is the hinge line for the principal plane. Before we close the principal plane onto the

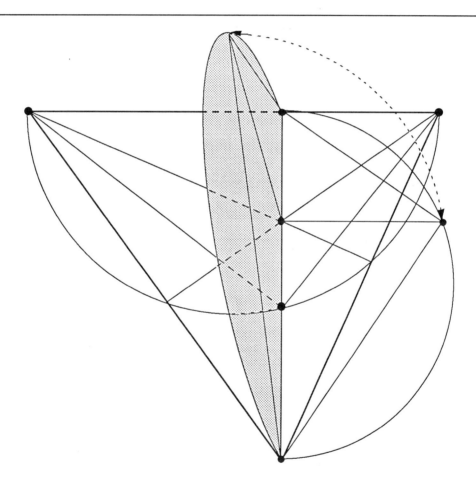

Figure 4-2. Door of Principal Plane

is the distance from the **pp** to the lens rear nodal point. Graphically, the lens rear nodal point is considered the camera station (**CS**). The effective focal length (coincident with the principal ray) is also called the principal distance, and is perpendicular to the image plane at the principal point. The gravity line, or vertical line, from the camera station intersects with the plane of the image at the nadir point. (The nadir point does not have to be within the visual image plane.) The line through the **pp** and the nadir point is the principal line. The point at which the principal line intersects the XY true horizon line is the **TH** point. The three points **TH**, **CS**, and **pp**, form a triangle with side **pp-CS** being the effective focal length. The triangle is a right triangle with the ninety degree angle at **pp.** There are two more right triangles that can be described in theprincipal plane that are of concern to us. There is a right triangle using the points **CS**, **pp**, and **VPZ**, with the ninety degree angle at **pp**. The line CS-**pp** is still the effective focal length. These two right triangles form another right triangle (**TH**, **CS**, and **VPZ**) with the ninety degree angle being located at **CS**. Graphically the line from **TH** to **VPZ** is the diameter for a semicircle which passes through **CS**. All of these geometric relationships are shown in Figure 4-2, but were not labeled. When labels have been applied, the geometry will appear as in Figure 4-3.

A lot of information was given in the preceding paragraph, so please study the paragraph carefully, and locate each geometric form described in Figure 4-3.

In the description of the principal plane and the focal length it may have seemed that there is one vertical plane of the focal length to consider. That is not true. Actually, we must look at the perspective pyramid (tetrahedron) shown in Figure 4-4. This perspective pyramid is made up of four triangles. The base triangle is coincident with the image plane and has as vertices the three major vanishing points (X, Y, and Z). The sides of the perspective pyramid are the three triangles with a common vertex at the **CS**. (The **CS** is the top of the perspective pyramid - tetrahedron.) Each side triangle is a right triangle with the ninety degree angle at the **CS** point. Then another set of triangles is formed using a point (**TH**) at the perpendicular bi-sector of each side of the base triangle, the opposite vertices (**VPX**, **VPY**, or **VPZ**), and the **CS**. There is more to be said in describing the perspective pyramid, the various triangles formed by it and within it; however we shall concern ourselves with only the two triangles, **VPX-CS-VPY** and **TH-CS-VPZ**, in connection with the focal length. It should also be noted that the perspective pyramid, as shown, is present only in three-point perspective. The pyramid becomes totally distorted in one- and two-point perspective, and one or more of the vertices are pulled out to infinity.

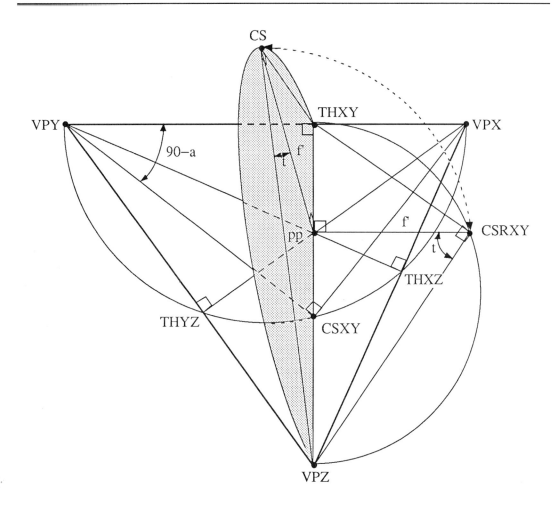

Figure 4-3. Plane of Focal Length

When the two triangles **VPX-CS-VPY** and **TH-CS-VPZ** are rotated into the image plane (as shown in Figure 4-5) more unique geometric patterns become available. In two-point perspective, the **TH-CS-VPZ** triangle is not available because **VPZ** is at infinity. The triangle **VPX-CS-VPY** is available and the distance **TH-CS** is the focal length. Therefore all the photogrammetrist has to do is to locate the camera station to find the effective focal length.

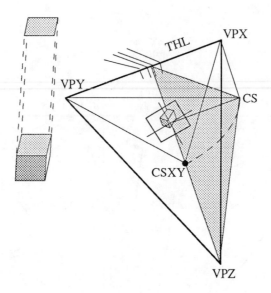

Figure 4-4. Perspective Pyramid

4.2 Rotation Angles

There are three rotation angles to consider in single-photo perspective - azimuth (**a**), tilt (**t**), and swing (**s**). In brief terms the rotations are about the Z,x,z axes. The rotation that relates the position of the object-space X and Y axes on the horizontal reference plane is **a**, a rotation about the Z axis. The rotation of the camera relative to nadar is **t**, and is the camera rotation about an axis perpendicular to the principal plane. The **s** rotation of the camera is about the principal ray, and is a rotation about the image-space z axis.

One-point perspective rotations are easily defined because the object-space planes are parallel to the image-space plane. The azimuth angle is either 0° or 90°, depending upon whether the object space plane is the YZ plane or the XZ plane. The tilt angle is 90°, placing the principal*on the horizon line. Thus, the swing angle is the only angle in one-point perspective that is variable. By definition, the swing angle can vary from 90° 180° to 270°, and is the relationship of the camera format with the horizontal plane. When the bottom edge of the format is parallel to the horizontal plane the swing angle (**s**) is 180°.

* point

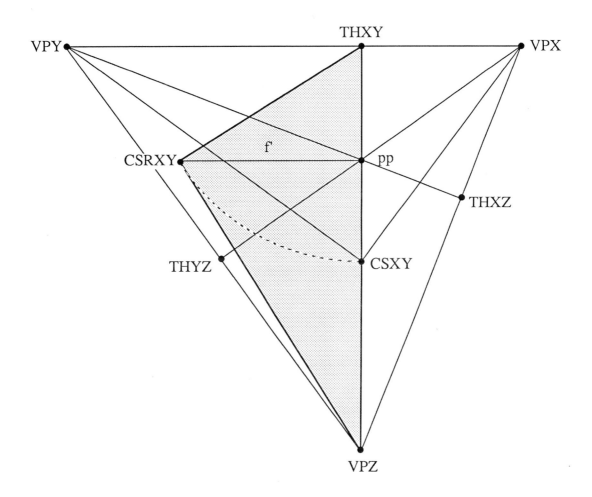

Figure 4-5. Triangles *VPX-CS-VPY* and *TH-CS-VPZ* Closed onto the Image Plane

In two-point perspective there are only two angles to determine, because the tilt angle is 90° by definition. Again, the swing angle can vary from 180° to 270°. The azimuth angle (**a**) is an orthographic angle in that it describes the rotation of the vertical plane of the object away from the orthographic vertical plane - a rotation about the Z axis. Remember, the XY horizon line is the intersection of two planes: the plane of the image, and the horizontal plane through the camera station. In orthographic projection, the trace of the XY horizon line on the horizon plane is also the intersection of a vertical plane (the image plane) with the horizon plane. The azimuth angle is the rotation of the orthogonal object away from this vertical plane (see Figure 4-6).

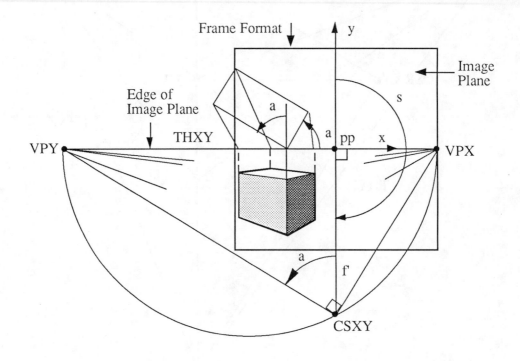

Figure 4-6. Two-Point Perspective Rotation Angles

Three-point perspective rotation angles are the same angles, but the tilt angle is never 90°, although the swing angle can very easily be 180°. In all cases of perspective the rotation angles describe the rotated relationship of the object, the image plane, and associated orthographic planes.

4.3 Conclusion

In this chapter we have discussed the effective focal length (**f'**) and rotation angles azimuth (**a**), tilt (**t**), and swing (**s**), and how they can be graphically determined from perspective imagery. To those just getting acquainted with perspective geometry and photographic images, the concept of a changing focal length may be something new. Cameras typically used in the collection of close-range photography do not have a fixed lens - the lenses are adjustable to enable the user to sharply focus the object-space image on the film. The focal length at the instant these collections are made is the **f'**, and it will be larger than the manufacturer's defined nominal focal length. A good rule of thumb, for graphical solutions is to approximate the **f'** by increasing the nominal focal length by about 2 percent. The perspective pyramid, or tetrahedron, was used to show the relationship to the three-dimensional image space of the camera to the two-dimensional space of the image. This relationship not only portrayed the **f'**, but also described the relationship of the three rotation angles. These rotations that relate the position of the object-space X and Y axes on the horizontal reference plane (**a**, a rotation about the Z axis), the rotation of the camera relative to nadir (**t**, a rotation about an axis perpendicular to the principal plane), and the rotation of the camera about the principal ray (**s**, a rotation about the image-space z axis). These parametric values - **f'**, **a**, **t**, **s** - are Phase One parameters, and will be used in the graphical and analytical analysis of single-photo perspective to develop the orthographic views.

DIMENSIONAL ANALYSIS THROUGH PERSPECTIVE

Chapter Five
One-Point Perspective: Step-by-Step Analysis

In this chapter we will discuss the perspective geometry of one-point perspective and some methods for using it to reconstruct object-space dimensions. Foremost in the one-point perspective methods will be projecting scale from parallel plane to parallel plane, and the squaring up (proportion division, or Canadian Grid), method of dimensional analysis. These methods use combined graphical and analytical procedures.

5.1 One-Point Perspective or Orthographic Geometry

One-point perspective is unique in that for all intents and purposes the image is an orthographic image (see Chapter Two, Section 2.3.5). All lines in a one-point perspective plane are true ratio lines relative to all other lines in the plane, provided that the object-space plane is parallel to the image plane. With that statement as a start, let's digress and define some of the parametric relationships and basic procedures involved. It is a geometric property of one-point perspective that, for any object-space plane parallel to the image plane, all the geometric shapes in the object-space plane remain the same in the image plane. (This relationship is best found in the study of optics.) In particular, all distances in such a plane are rendered in true ratio, and hence are imaged orthographically. This orthographic image will either be a true elevation or plan view, or it will be defined as an auxiliary orthographic view. (An auxiliary view is any view not defined as a plan or elevation view - French, ~~1978~~ 1957.)

5.1.1 Standard Parameters

The standard single photo angular (rotation) values have a fixed relationship. The azimuth angle (**a**) is 0°, or 90° (important when the major coordinate plane — YZ or XZ — must be considered), the tilt angle (**t**) is 90°, and the swing angle (**s**) ranges from 90° to 270°. For all practical purposes, these relationships apply to an orthographic drawing (or blueprint) of the object-space plane. In Chapter Two we mentioned that a true-horizon-line (THL) has a definite purpose as the line between two of the major axis vanishing points (edge view of a major coordinate plane.) In one-point perspective the camera is looking along one of those major axes and the THL is still defined as parallel to the horizontal plane, or it can become a user-defined horizontal line through the principal point (vanishing point).

The effective focal length is not graphically defined, however it can be determined through the use of the scale (S) and range (R) calculations: $f' = SR$. It is also typical that the range is not provided. However, if we should have scale in two parallel planes (S1 and S2) and the distance between the planes (dR, as in delta range), it is possible to determine R. With R known, f' can be determined. An example would be a box of known length, width, and height. The box is parallel to the image plane. Since we know dimensions in the planes of the box parallel to the image plane we can determine the scale for each plane. Using $S1 = f'/R$ and $S2 = f'/(R + dR)$, where dR is the known length or width, the equation $R = (S2\ dR)/(S1-S2)$ is developed. The values of R are used in either scale equation to determine f'. Of course there is always 'Murphy,' who will not allow us to have a known box in the image. In that case you may allow yourself to be stumped, or you can provide yourself with a SWAG (scientific wild alternate guess) for the size of your box. You then continue, remembering that you are now developing model space values, and not object space values.

5.1.2 Graphical Layout

The graphical solution in one-point perspective is usually constructed directly over the photographic image. Although it is possible to project the working scale plane outside the image area, there is a greater potential for human error when using this drafting technique. The decision for doing this should be based on the complexity of the object whose dimensions are to be determined. With a more complex object, it is better to complete the graphical solution directly over the image (auxiliary view). It is convenient to use a plastic or acetate overlay in this application, and to replace the overlay periodically to avoid clutter.

Having defined these relationships, we will now describe examples of one-point perspective.

5.1.3 Proportional Division

It is possible to locate an arbitrary point in object-space by subdividing a rectangle in the coordinate plane of a one-point perspective image. Subdivide a rectangle (with corners labeled A, B, C, and D) in halves by drawing the diagonals AC and BD (locating the center, point E), then draw a line parallel to the sides AB and DC through point E, and label the new intersection points F and G. Point F is on the line AD, and point G is on the line BC. Two new rectangles, ABGF and FGCD have been created. This procedure, called proportional division, can be repeated so that each successive rectangle can be halved, thus producing sections equal to 1/2, 1/4, 1/8, 1/16, and so on, of the original rectangle (see Figure 5-1).

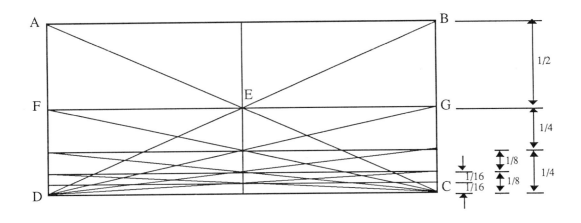

Figure 5-1. Proportional Dimensions of a Rectangle by Halves

Proportional schemes can, of course, be established based on fractions other than 1/2. Another proportional method is to divide the rectangle ABCD into thirds, and hence to produce sections equal to 1/3, 1/9, 1/27, and so on (C. P. Kelley, 1978-1985). The procedures for doing the one-third method is first to divide the rectangle in half (see Figure 5-2) creating points F and G (do not draw the line FG). Then draw the lines AG, BF, FC, and GD. Mark the intersections of the main rectangle diagonals (AC and BD) and these new lines, and draw lines through these points parallel to the lines AB and DC — these are the one-third lines (see Figure 5-2).

From these two examples (halving and thirds), it is easy to generalize to show that a rectangle can be divided into any number of proportional parts. First, suppose that

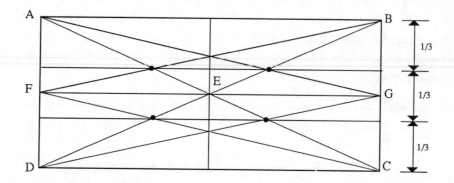

Figure 5-2. Proportional Division of a Dectangle by Thirds

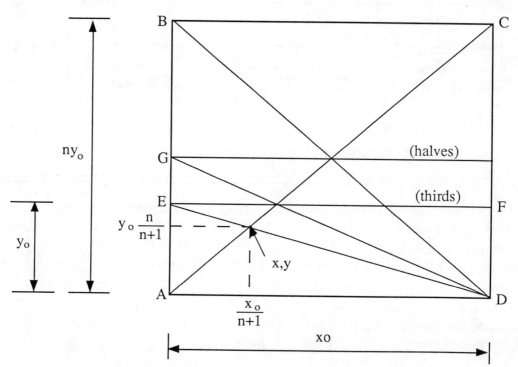

Figure 5-3. Horizontal Division of a Rectangle into n+1 Equal Parts, given Vertical Division of the Rectangle into n Equal Parts

proportional division into n parts has already been done in the vertical direction (see Figure 5-3). The rectangle ABCD has area n times rectangle AEFD. Draw diagonals AC and ED. The x coordinate of the intersection is 1/(n+1) of the way across the rectangle. To see this, write the equations of the intersection point (x,y) in terms of the lines generated by the diagonals: $y = n(y_o/x_o)x$, and $y = -(y_o/x_o)x + y_o$; expanding the right-hand side gives $(x/x_o) = 1/(n+1)$.

Another method for dividing a rectangle, shown in Figure 5-4, is to draw an arbitrary line BH, walk a compass along it n times, connect the last hatch-mark on BH with C, and then draw parallels to HC through all of the intervening hatch-marks. This method is more efficient than the diagonal method when the rectangle is imaged orthogonally, but cannot be applied otherwise. In contrast, the diagonal method can be applied to any rectangle, even when it is imaged in perspective, because distances and angles are not used in the method. Of course, for perspective rectangles, vanishing lines replace parallel lines in the construction. By using these methods of proportionality it is possible to transfer scale to any line in the plane of the rectangle.

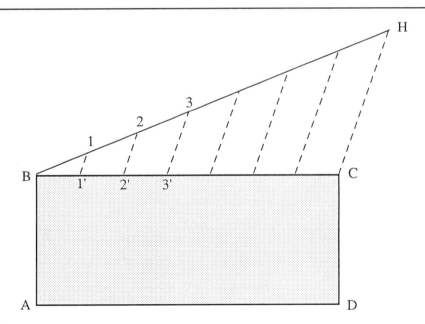

Figure 5-4. Horizontal Division of a Rectangle into Equal Parts using an Arbitrary Line with Equal Divisions

5.2 Basic Example of One-Point Perspective

To illustrate the basics of one-point perspective, imagine you have set up a camera at one end of a hall. The camera is in the center of the hall, pointed directly down the center line towards the end wall. The image plane (film plane) of the camera is parallel to the end wall (**t** = 90°). Also, the fact that the image of the end wall forms a true rectangle is a good indicator that the geometry is one point perspective. (Murphy can fool us with an Ames window geometry set: a trapezoid imaged as a rectangle - Cohen, ~~1982~~) The end wall contains a door and a window. When the photograph is developed the image will be one-point perspective, and all lines, angles, and shapes in the plane of the end wall will be in true-view relationship. You will also see that the lines along the length of the hall (including the tops of doorways) can be extended to an intersection point, the one vanishing point. It is also true that the principal line (principal ray) is perpendicular to the object plane (the end wall), and therefore is parallel to the intersection lines (vanishing lines). Because the principal ray is parallel to the lines defining the vanishing point, the vanishing point and the principal point are the same. This is proof for one-point perspective geometry, and in the graphical analysis a small angular shift (plus or minus 3°) of the principal ray away from the center of format will not be significant. The geometry of this illustration is shown in Figure 5-5. Thus it can be said that, if all the shapes on the image of the end wall remain proportional to the actual shapes, the image plane of the end wall is an exact replica of the object-space plane, and that portion of the photograph is actually an orthographic view. All the facts cited in this example are important because they are always true in one-point perspective. This example has been worked and is in the Appendix (A-5) as Example One (E-1).

The apparent paradox of a perspective image that provides an orthographic view can be explained by the fact that there is constant scale across the perspective plane. Scale can be defined as the effective focal length (**f'**) divided by the range (**R**). In close-range photography the lens is focused at a finite distance (not at infinity), and for each change of focus there is a related **f'**. (A discussion of focal length geometry is found in Chapter Four, and this relationship was discussed in the beginning of this chapter.) The finite distance is **R**, or the distance from the camera lens to the point of focus, in the depth of field for focus. It may seem that the scale at the edge of the photograph should be different than the scale at the principal point because the **f'/R** relationship is defined for only the principal ray. However, the scale for every other point in the photograph is defined using the proportional geometry of the visual ray from the image point through

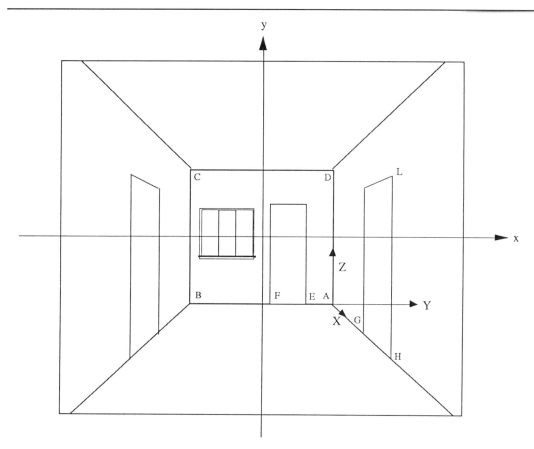

Figure 5-5. Hallway Illustration of One-Point Perspective

the lens to the corresponding object points on a parallel object-space plane. The proportionality is constant for all of the visual rays of the image, and for all practical purposes the scale relationship remains constant for every imaged point of the parallel object-space plane. This means that scale is the same over independent planes of the one-point perspective.

Now let's open the door in the one-point perspective plane, and include the new wall across the room in our photograph. This new wall will also be a one-point perspective plane, but it will have a smaller scale because the range (distance) from the camera is greater. The question is, "How do we determine this new scale?" The answer is simple.

Using the vanishing point (principal point) we project the door opening to the new wall, and use the known or computed dimensions of the projected door to define scale on this new one-point perspective plane. Although we said the answer is simple, there is a requirement. The requirement is that the image must contain the intersection line of the floor and the new wall. Example Two (E-2) in A-5 describes this method. When we project the door opening, we are actually projecting the corners of the plane of the door opening, as parallel lines, to the vanishing point. It is impossible to determine the intersection of the projection of the top two corners of the door plane and the new wall without some **a priori** knowledge of the point of intersection. We obtain this knowledge with the help of the floor plane. The floor plane is common to both vertical planes (door and wall). This is illustrated in Figure 5-6. Without this **a priori** knowledge, it would not be possible to complete the projection, and hence transfer scale to the new wall; e.g., if a piece of furniture were placed so that the new wall and floor intersection is not in the photograph, the projection would not be possible.

Likewise, we can now determine the distance from the plane of the door to the new wall. Remember, we are working with one photograph. The scale is defined as f'/R, and we have a scale for the plane of the door, SD, and a scale for the wall, SW (same procedure used in 5.1.1). The two scale equations can be written as $SD = f'/RD$ and $SW = f'/RW$. Since f' is the same in both equations, the equation for the range to the wall can be written as $RW = SD \, RD / SW$. As an aside, there is an old saying that the reason we see magic is that we don't know the trick. The more we state that something is impossible without **a priori** information, the more we seem to limit ourselves to giving a small number of solutions. Thus, if readers limit themselves to just the examples given in this manual, they are more likely to see magic, rather than the trick. (Remember, each photograph, even from the same roll of film, is unique.) It is our intent that this manual will be used to help look for and find the trick. For example, we have already stated that orthographic drawings (plan and elevation views) can be used to construct a perspective drawing, not unlike a photograph. When working with photographed man-made structures, it is likely that orthographic drawings (blueprints, floor plans, and so on) are available for the overview of a particular plane.

The graphical solution, as far as it can be developed, is obtained at the same scale as this additional information. This is another possible solution to the door in the end wall and the new wall beyond it. The two views are overlaid, and useful information is projected from the orthographic drawing to the perspective lines of the photograph

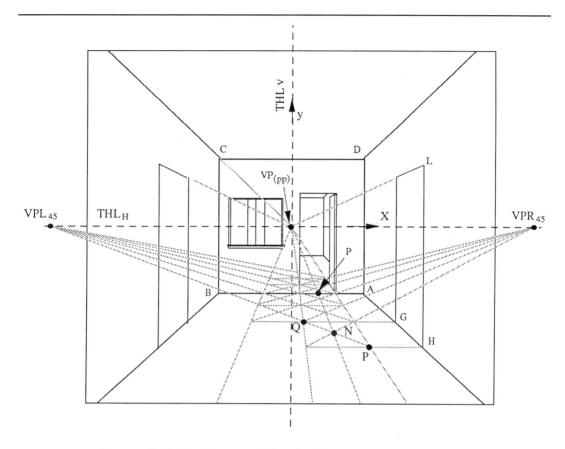

Figure 5-6. End-wall of Hall in One-Point Perspective

to provide intersecting planes and lines that might not otherwise be available. The dimensions of items in the plane of the new wall can then be determined. When working with the geometry of single-photograph perspective, the photogrammetrist must expect to be resourceful in assimilating all available information.

The illustration of the hall has provided the basic concepts required. Now let's consider a cropped photograph of an airplane hangar. The hangar is imaged as one-point perspective so that we can see through the structure because the front and back hangar doors are open. The camera was not positioned on the center of the hangar (see Figure 5-7). The camera was positioned to the right of center of the hangar, and directly in front of the camera (between the camera and corner of the hangar) is a flat-bed cargo cart. On the cart is a crate (parallel to the image plane), and we need to determine the length, width, and height of the crate. The only known dimension available is the height of a utility crane behind the hangar, but visible through the open doors. The floor of the hangar and the apron before and behind the hangar are all on the same horizontal plane. The solution will use procedures that are similar to those used with the hall, the open door, and the new wall. This is illustrated and described as Example E-3 in the Appendix.

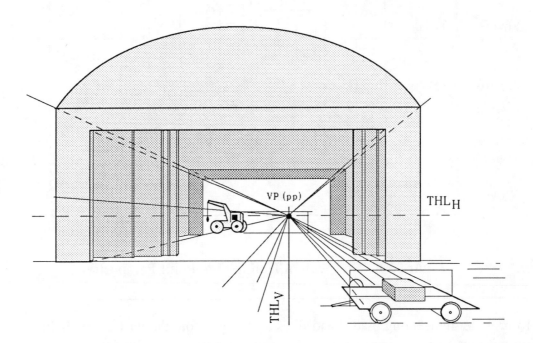

Figure 5-7. One-Point Perspective - Hangar and Cart

The lines of the hangar, the cargo cart, and the crate are used to determine the vanishing point. Horizontal lines are used to establish a true horizon line (THL). The known height of the utility crane is drawn so that the line is perpendicular to the THL. The intersection of the known vertical and the horizontal plane is determined. The known vertical is then projected to the front plane of the hangar. Lines are drawn to project the front and back of the crate to the horizontal plane under the cargo cart. These lines in the horizontal plane are then projected to the front plane of the hangar. The length and height are now determined. With this information it is now possible to determine the scales for the front and back planes of the crate. With these scales it is possible to determine the width of the crate.

This last example illustrates more than how one-point projections are used in determining dimensions. It is also an example of the importance of good drafting techniques. All lines and projections are drawn, and therefore are susceptible to human error. The photogrammetrist must utilize a certain drafting skill to eliminate potential human errors.

5.3 Conclusion

The main facts for one-point perspective are: azimuth is 0° or 90°, tilt is 90°, object plane parallels image plane, principal ray and vanishing lines are parallel, and the vanishing point is the principal point. These facts are also another way of stating that every object-space plane parallel to the image plane, in the depth of field for focus, has a unique scale that is constant in that plane. Also, if you know one dimension in one of those object-space planes, you can determine the scale in all of the planes parallel to the image. With a scale known in these planes it is possible to determine the distance between object-space planes. Thus, with **a priori** knowledge and some drafting skill, it is possible to use the geometry of one-point perspective to obtain the dimensions between horizontal and vertical planes. This information is used to determine the dimensions of the target areas visible in the photograph.

DIMENSIONAL ANALYSIS THROUGH PERSPECTIVE

Chapter Six
Two-Point Perspective: Step-by-Step Analysis

6.1   Introduction

In this chapter we discuss some illustrative methods useful in reconstructing three-dimensional geometry from two-point perspective photography (images). For two-point perspective, we will show that **a priori** knowledge of a single diagonal angle is enough to reconstruct a scale model of the solid. The coordinates of two object-space points or one object-space dimension will be required to provide scale. In the process of constructing the object-space model we will determine the vanishing points, the camera station (**CS** in the image and in object space), the perspective principal point (**pp**), the perspective focal length (**f'**), and the camera rotation angles of azimuth (**a**), tilt (**t**) and swing (**s**). By using the lens center as a projection point, we reconstruct three true orthographic views of the object on paper. In contrast with other methods (Gracie et al., 1967; Busby, 1981; and Novak, 1986), this reconstruction will not require a full-format image. These methods are adapted from geometrical procedures used to produce architectural drawings (McCartney, 1963; Walters and Bromham, 1970), but of course the goal is to infer the three-dimensional geometry from an image rather than to create the image from three-dimensional information.

6.2   The Fundamentals of Graphical Two-Point Perspective

The most basic construction in our two-point perspective analysis involves parallel lines on the horizontal plane defined by the imaged rectangular solid. These parallel lines converge to vanishing points (see Figure 6-1). From this vanishing point construction evolves the principal point, the focal length, and the camera station of the perspective image-space model (Gracie et al., 1967; Kelley, 1978; Slama, 1980; Moffitt and Mikhail, 1980; Wolf, 1983).   *1978-83*

1974

6.2.1 Perspective Comparisons

To grasp the concept of two-point perspective, it is important to compare the geometry of the graphical image-space models of two- and three-point perspective. The edge views for two- and three-point perspective image-space models are shown in Figure 6-2a and 6-2b. When the tilt (**t**) is 90° (Figure 6-2b) the image plane becomes

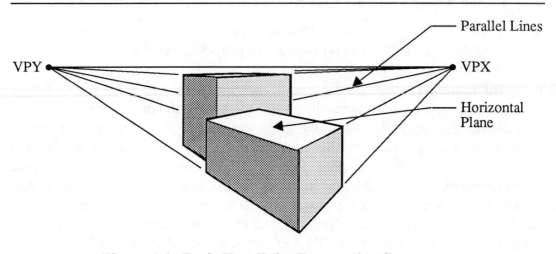

Figure 6-1. Basic Two-Point Perspective Geometry

parallel to the nadir line (the line **CS-VPZ**); the point **VPZ** goes to infinity; and the points **VPX, VPY**, and **pp** become coincident with the true horizon line (**THL**). Note that, by definition, **THL** is the trace of the horizontal plane, through the camera station, across the image plane. As shown in Figure 6-2c, the line between **VPX** and **VPY** (labeled **THL**) is the diameter of a circle and the **CS** is a point on the circle. To illustrate this graphically while keeping the relationships of the geometry true, the semicircle from **VPX** to the **VPY** through the **CS** is rotated 90° about the **THL** until the semicircle lies in the image plane. True relationships that were constructed in Figure 6-2b (the side view) may now be shown in Figure 6-2c (the front view). In fact, any image model vertical relationships that can be constructed between **CS** and the image-space model **THL** may still be shown in the rotated horizontal plane of Figure 6-2c, as if the horizontal plane had not been rotated. This is a standard procedure of descriptive geometry and is very practical in graphical photogrammetric solutions.

6.2.2 Vertical Lines

In our two-point perspective analysis (see Figure 6-3), parallel planes, such as ABCD and EFGH (top and bottom of the rectangular solid) are defined to be horizontal. The edges of the base — HG and HE — define the X and Y axis directions, and point to the associated vanishing points **VPX** and **VPY** of the image. Between the vanishing

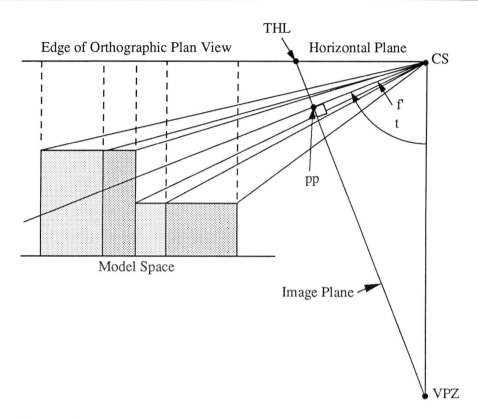

Figure 6-2a. Image-Plane Edge-View of Three-Point Perspective

points is the **THL**. Because the image plane is perpendicular to the object horizontal planes, all vertical lines of the object are parallel to the image plane, and the principal ray points directly at the **THL** — defining a tilt of 90°. (Actually, two-point graphical procedures are usable whenever the tilt is 90°, plus or minus 3°.) Thus, the vertical imaged lines have no graphically definable vanishing point, because such a point would be an infinite distance above or below the **THL**.

6.2.3 Diagonal Lines

If a two-point perspective image has full format, its perspective principal point (**pp**) may best be estimated as the image's center of format (center of vision, see Busby,

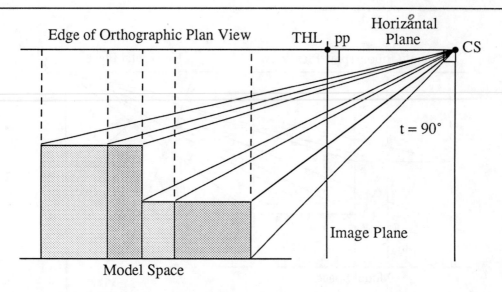

Figure 6-2b. Image-Plane Edge-View of Two-Point Perspective

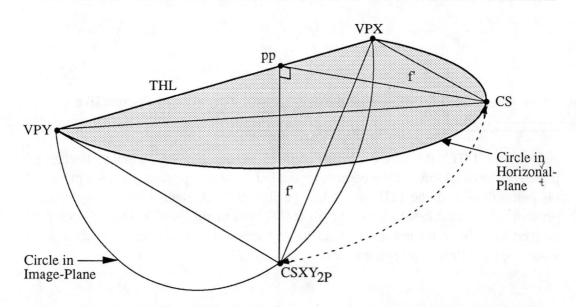

Figure 6-2c. Two-Point Perspective - Rotation of Image Plane into Horizontal Plane

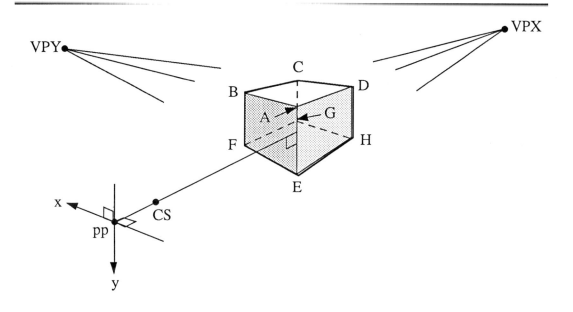

Figure 6-3. Two-Point Perspective and Vertical Line Illustration

1981; image-center, see Novak, 1986). However, even without full format, a two-point perspective image allows three-dimensional reconstruction given some other **a priori** knowledge. The reconstruction starts by determining the location of the camera station (**CS**), and thereby the **pp** and effective focal length (**f'**). The **CS** can be found by locating the vanishing point of horizontal or vertical-plane diagonals, given a known angle (**thetah** or **thetav**) of the diagonal in object space (see Figures 6-4 and 6-5). The easiest diagonal to use is the diagonal of a horizontal or vertical square enclosing a circle. A manhole cover or an automobile wheel rim are such circles and can graphically (using **VPX** and **VPY**) be enclosed in a square whose angle is known.

6.2.4 Six-Step Procedure - Horizontal Diagonals

A procedure for using a horizontal-plane diagonal to locate the **CS** and **pp** is illustrated in Figure 6-4. There are six steps in this procedure. Using any horizontal plane (with known diagonal angle) on the object, such as ABCD in Figure 6-4, the first step is to locate the vanishing points **VPX** and **VPY** for the X and Y coordinate axes.

The second step is to draw the **THL** between the two vanishing points and a semicircle with the **THL** as the diameter. With two-point perspective it is best to draw the semicircle beneath the **THL**, although the solution can be worked with the semicircle above the **THL**. Third, locate the intersection point of the **THL** with the horizontal-plane diagonal line (AC) extended. This is the vanishing point for the horizontal-plane diagonal (**VPDH**). Fourth, on a clear piece of overlay scribe two lines at right angles and between them the known-angle line   ($\theta = 45°$ for our square). The graphical intersection of these three lines (each passing through a vanishing point) on the semicircle will be the **CS**. Each line must be long enough to extend beyond its respective vanishing point, allowing easy traverse of the semicircle. Fifth, move the position of the **CS** around the semicircle, and, at the instant when each line simultaneously passes through a vanishing point, you have located the perspective image position of the **CS**. Sixth, draw a line perpendicular to the **THL** and through **CS**. The intersection of the line with the **THL** locates the **pp** and the **f'**. This completes the six-step procedure.

   The construction shown in Figure 6-4 can be understood geometrically by rotating the semicircle (with **CS**) about the axis **THL** 90° out of the paper (image plane, see Figure 6-2c). **CS** now represents the true (three-dimensional) camera station location and **pp** is the true principal point as drawn. |**Note:** The last phrase "as drawn" is

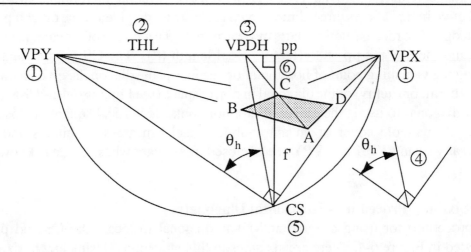

Figure 6-4. Six Steps using Horizontal-plane Diagonals

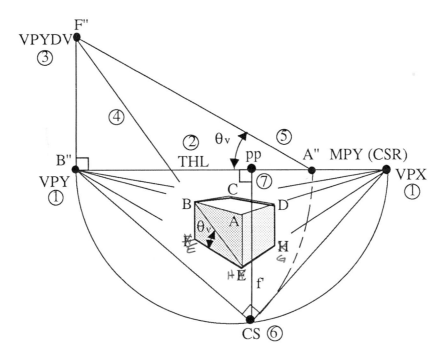

Figure 6-5. Seven Steps using Vertical-plane Diagonals

a very important statement. This implies another "oh, by the way." Remember that the true location is relative to the solution model or relative to the scale of the graphical solution.| The line **CS-VPY** is parallel to the true line AB for the horizontal rectangle, and **CS-VPX** is parallel to the true line AD. The line **CS-VPDH**, at angle **thetah** from the line **CS-VPY**, is parallel to the true diagonal AC, hence the vanishing-point interpretation of **VPDH**.

6.2.5 Seven-Step Procedure - Vertical Diagonals

Equally straightforward (see Figure 6-5) is the use of a vertical-plane diagonal to locate the **CS** and **pp**. The first two steps are the same: locate the X and Y vanishing points (**VPX** and **VPY**) and draw the semicircle below the **THL**. The third step is to draw a perpendicular to the true horizon from either the X or Y vanishing point, depending upon whether the vertical diagonal is in the YZ or XZ plane. In Figure 6-5, the extended diagonal (HB) is in the YZ plane and passes over the Y vanishing point, so the perpendicular is through **VPY**. Fourth, extend the diagonal to intersect with the perpendicular just constructed. The intersection is the vanishing point for the vertical diagonal (**VPYDV**). Fifth, find the position on the **THL** so that lines to **VPYDV** and **VPY** subtends the known angle **thetav** relative to this point (in the case of Figure 6-5, **thetav** is 30°). The point so located is normally referred to as a Y measuring point (**MPY**), but it is also the camera station revolved (**CSR**). The point **VPY** is the center of revolution for constructing the **CS** from the **CSR**. Sixth, using **VPY** draw an arc from **MPY** that intersects with the semicircle. The intersection point is the **CS** (model camera station). Step seven is to draw a perpendicular from the **THL** through **CS**, thus locating **pp** on the **THL**. This completes the seven-step procedure.

The construction of Figure 6-5 can be understood geometrically (after the fact) by rotating the semicircle (with **CS**) about **THL**, 90° out of the paper (image plane). As with Figure 6-4, **CS** and **pp** are thereby tendered in accurate three dimensional positions. Because the diagonal HB is in the same plane as HE and AB, and this plane is vertical, it follows that the vanishing point for HB is in the image-vertical direction from **VPY** (the vanishing point of HE and AB). Hence, the vanishing point for HB is found by extending HB to intersect (at **VPYDV**) the perpendicular line through **VPY**, as illustrated. With the camera station in its true image-model three-dimensional position (after the rotation described at the beginning of this paragraph), the line from true-**CS** to **VPY** and the line from true-**CS** to **VPYDV** make the same angle **thetav** as the lines HE and HB on the object. This is because a vanishing point is a meeting of the imaged parallel lines of object space, hence, true **CS**-**VPY** is parallel to true HE, and true **CS**-**VPYDV** is parallel to true HB. To depict these lines graphically (from true **CS** with the angle **thetav**), rotate the true CS about **VPY**-**VPYDV** until these lines lie in the plane of **THL** and **VPYDV** (the image plane). The point on the **THL** now occupied by **CS** is the camera station revolved (**CSR** or measuring point Y — **MPY**). This location of **CSR** is exactly the same as that obtained by rotating the **CS** in the image plane about **VPY** to **THL** (as shown in Figure 6-5). The converse is true when the location of MPY on the **THL** is known (having been located by knowing **VPYDV** and **thetav**). The image-plane **CS** is located by rotating **MPY** about **VPY**, until **CS** is generated at the intersection with the semicircle. A perpendicular from the **THL** through this newly located **CS** determines the **pp** and the **f'**.

6.3 Accuracy

To support accuracy for the solution, all available parallel diagonal lines should be used to locate the horizontal- or vertical-plane diagonal vanishing point. If possible, at least three lines should be used. In fact, if it is possible to use both a horizontal- and vertical-plane diagonal vanishing point to locate the camera station, it would establish a greater confidence level in the solution than if just one diagonal vanishing point were used.

6.4 Rotation Angles

Once the construction of **pp** is complete, the **f'** is then measurable as the distance between **pp** and **CS**. Also, of the three rotation angles (azimuth, tilt and swing, as defined by Slama, 1980), the azimuth angle (**a**) is the angle between the **THL** and the line from the **VPY** to **CS**. All of these parameters are identified in Figure 6-6.

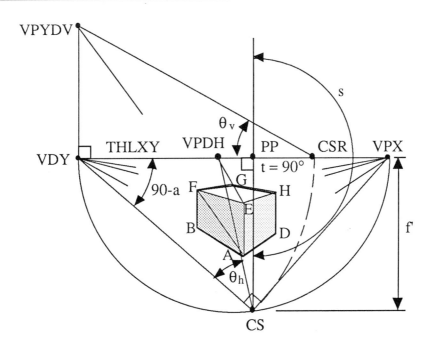

Figure 6-6. Two-Point Perspective Parameters Identified

This leaves the swing angle (**s**) to be determined. The swing angle defines the rotation of the image frame relative to the true horizon line. By definition (Slama, 1980), swing angle is "the angle at the principal point of a photograph which is measured clock-wise from the positive y-axis to the principal line at the nadir point." The image coordinate system is defined with its x-axis parallel to the bottom format edge, with its y-axis perpendicular to the x-axis, and with its origin at the principal point. If the camera was level to the horizontal plane of the object-space coordinate system at the time of exposure, the swing angle will be 180°. If the camera was not level to the horizontal plane and the image had full format, then the swing could easily be measured. If there is no image full format, swing (shown in Figure 6-6 for definition and clarity) is always taken to be 180°, this choice having no consequences in graphical solution.

6.5 Plan and Elevation Views

Given **CS**, **VPX**, and **VPY**, it is possible to draw a true-ratio plan view and elevation views of the imaged object. The plan-view construction is shown in Figure 6-7. Draw perpendiculars to the **THL** from the object as shown. Using point A' (which is the same as A" because the vertical reference plane is **THL** in this example) as one corner of the plan view, draw lines parallel to **CS** - **VPX** and **CS-VPY** through point A'. Extend lines from **CS** through D' and B' until they intersect the newly drawn lines at D" and B", respectively. Lines are drawn perpendicular to A' D" and A' B" through the points D" and B" to intersect on the extended lines **CS-C'**. This forms the corner C". The accuracy of the construction is visually checked at point C" by how closely the intersections of the three lines cluster. Adjustments may be made to the graphical solution until the intersection at point C" is acceptable.

6.5.1 Plan View

The construction in Figure 6-7 is obtained by rotating the plane of the image semicircle into the plane of the plan view, projecting vertical lines to the **THL** (or plane hinge) and drawing lines from the projection center (**CS**) through the points on the **THL**. (Note that the **THL** is also an edge view of the image plane). These extended lines intersect the XZ and YZ planes and locate the position of the vertical lines as points in the plan view. The angles of A'B" and A'D" with respect to the **THL** are clearly the same as the angles between the **THL** and the **CS-VPY** and **CS-VPX** lines, respectively. In this scaled plan view, with the origin of the object space coordinate system at point A', the camera station coordinates Xc, Yc are easily found. Note that the object-space plan view is arbitrarily placed with respect to the **THL**. The relation of the plan-view scale to the image scale is dictated by the positioning of an additional reference plane parallel to the image plane. In this example the image plane was used as the reference plane.

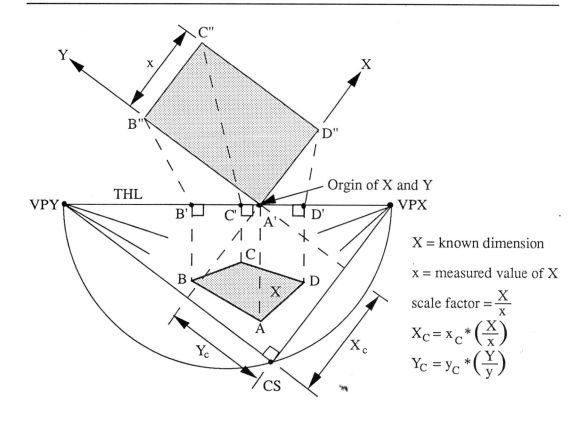

Figure 6-7. Two-point Perspective Plan View and CS Coordinates Xc and Yc Constructed to the Same Scale.

6.5.2 Elevation View

The construction of a true-ratio side view of a vertical plane of the rectangular solid (see Figure 6-8) is even easier, and also gives rise to the camera station coordinate Zc with the same scaling factor. Given **VPY**, **VPX**, and **CS**, extend line EA to meet **THL** at A'. Extend diagonal EB until it meets the vertical line of **VPY**. The intersection point is **VPYDV**. Rotate **CS** about **VPY** in the image until the intersection **MPY** (E") is located on **THL**. Then, the true length-to-width ratio of the vertical rectangle is generated by E", F", and B". The coordinate Zc also appears implicitly in Figure 6-8, for EA'/EA on the image is equal to Zc/(true EA) in object space. Hence, EA" represents Zc to the scale of EA. An elevation view of EFBA is obtained by completing the rectangle E"F"B"A' as shown. The construction can be understood in the same way as that of Figure 6-5, and can be used to construct the other elevation view of the rectangular solid by using **VPX** instead of **VPY**.

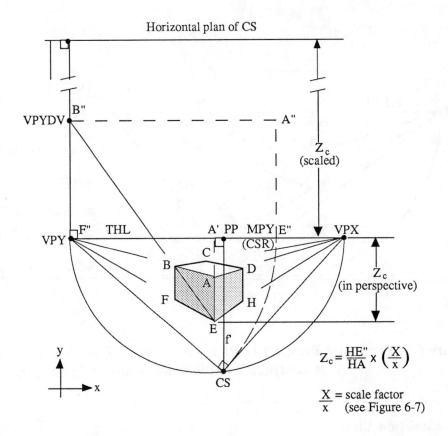

$$Z_c = \frac{HE''}{HA} \times \left(\frac{X}{x}\right)$$

$\frac{X}{x}$ = scale factor (see Figure 6-7)

Figure 6-8. Two-point Perspective True-ratio Side View and Construction CS Coordinate Zc Relative to HA.

6.6 Constant Scale

Although the two side views and the plan view obtained by these constructions are not to the same scale, they offer enough redundancy to be converted to the same scale (e.g., the scale for which AD=1cm). Of course, the true length AD is still undetermined by this single-image construction, unless a dimension (say, AD = X in Figure 6-7) is known. In that case, all the true distances are easily retrievable starting from the graphical scale views. Example E-4, in the Appendix, describes and illustrates the two-point procedures of this chapter.

If the imaged object is complicated, three-dimensional reconstruction may proceed piecemeal with both graphical and analytical methods. With complicated object-space geometry, many plan views and elevations can be constructed, each with its own scale; the scales are reconciled analytically by referring model dimensions to known object space dimensions. To acquire the true scale requires knowing the coordinates of at least two object-space points relative to the CS object-space coordinates, or knowing one dimension in a true-ratio view. Many times complicated solutions are completed using programmed analytical procedures.

6.7 Conclusion

This chapter clearly shows that when two-point perspective geometry exists in a single image, **a priori** geometric knowledge can enable graphical reconstruction of three-dimensional relationships from that imagery. The methods used to reconstruct three-dimensional relationships might seem contrived, but how many times in the past have single images, with no format and two-point perspective, gone unused because such a simple solution was not considered? In the absence of full format, we used a diagonal of known rotation to determine the camera station and therefore the perspective principal point, perspective focal length, and camera rotation angles. The method of diagonals is just one available for a graphical photogrammetric solution. There are many other methods, and of course each of them is only as accurate as the known values and the user's ability to work the solution. The choice of two-point perspective methods will always be defined by geometry cues in the imagery, and with man-made objects these cues are nearly always present.

DIMENSIONAL ANALYSIS THROUGH PERSPECTIVE

Chapter Seven
Three-Point Perspective: Step-by-Step Analysis

7.1 Introduction

In Chapter Six, graphical techniques for mensuration of two-point perspective images were discussed. However, it was necessary to know at least one diagonal angle of a rectangle to reconstruct the principal point and camera station of the perspective image. These in turn were used to reconstruct orthographic views of an imaged object. Whereas previous treatments of three-point perspective (e.g., Gracie, et. al., 1967; Busby, 1981; and Novak, 1986) required prior knowledge of the principal point of the image, the approach presented in this chapter (similar to that of Kelley, 1978-1983) requires no such knowledge.

This chapter extends the methods of the previous chapter to the problem of three-point perspective geometry (in which none of the edges of the imaged solid are parallel to the image plane). In particular, we consider a rectangular box (Example E-5 in the Appendix), and also solids with two nonrectangular horizontal surfaces (Example E-6 in the appendix). The methods prove to be somewhat simpler than in two-point perspective, and knowledge of a diagonal angle is no longer required. As in the previous chapter, the scaled camera position and attitude are determined together with the orthographic views of the solids.

7.2 Reconstructing a Rectangular Box

Reconstruction of orthographic views of a rectangular box is an easy example. This involves performing a graphical resection that locates the attitude and camera-station coordinates scaled with respect to an arbitrary object-space coordinate system. The arbitrary object-space coordinates system in this case is defined by the forwardmost corner of the imaged box.

7.2.1 Six-Step Procedure for Reconstruction

For our example (E-5) consider the three-point perspective image of a box in Figure 7-1. The top of the box (ABCD) can be reconstructed in the following six steps.

Step l. Construct the vanishing points by extending the three sets of parallel edges of the box so that they intersect. These constructed intersection points are called vanishing points, and are labeled VPX, VPY, and VPZ, according to the respective axes producing them.

Step 2. Locate the perspective principal point (**pp**) on the image as the intersection of the three altitudes of the triangle formed by VPX, VPY, and VPZ.

Step 3. Construct a circle whose diameter is the segment VPX-VPY; label as the camera-station position CSXY the intersection of this circle with the altitude drawn through VPZ.

Step 4. Construct lines from VPZ through A, B, C, and D that intersect the line VPX-VPY (the true XY horizon, or THLXY). These points are labeled as the respective points A', B', C', and D'.

Step 5. Construct a line through A' parallel to CSXY- VPY, and extend CSXY-D' to intersect this line at a point (D"). Similarly, construct a line through A' parallel to CSXY-VPX, and extend CSXY-B' to intersect this line at a point (B").

Step 6. Construct C" by completing the rectangle A'B"C"D". (A test of the drawing accuracy is the proximity of C" to the extended line CSXY-C'). This completes the six-step construction. The other two orthographic views of the solid can be obtained analogously. The orthographic view of face AEFB is constructed by projecting from VPY to THLXZ instead of from VPZ to THLXY (see Figure 7-1); face ADHE is reconstructed by projecting from VPX to THLYZ. Of course, the scales of the three orthographic views obtained in this way will probably be different. To refer the views to a common scale, identify the same segment (e.g., representing edge AB) in two views, and then change the scale of one view so the two representations of AB have the same length. This can be done graphically by parallel displacement of the relevant horizon lines (THLXY, THLYZ, or THLXZ).

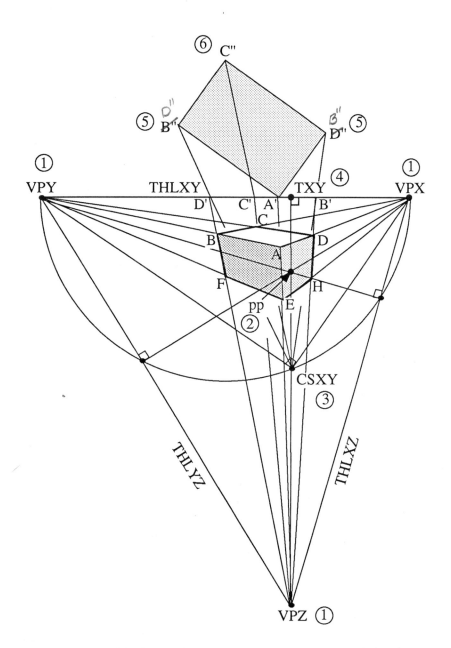

Figure 7-1. Six Steps to XY Plan View of Rectangular Box

7.3 A Geometric Justification of the Method

The construction shown in Figure 7-1 can be understood geometrically in much the same way as the construction in the two-point perspective discussion of Chapter Six.

First, we show the validity of the construction of CSXY and **pp** (the first three steps of the above method). The acquisition geometry of the image in three dimensions places the camera station (**CS**) one effective focal length away from the positive image plane (and object space) along the principal ray. Figure 7-2 illustrates an auxiliary view of this geometry and the vanishing points of the perspective image. Revolving **CS** about the true horizon line THLXY (to the image plane) generates the point CSXY. The standard perspective projection model implies that a ray from the three-dimensional **CS** through any image point (of a positive image) will intercept the corresponding object point. These rays may be considered visual rays. In this model, the constructed lines from the **CS** through the image vanishing points (VPX, VPY, VPZ) are parallel to edges of the object-space box. There is a tetrahedron formed by the image-space points VPX, VPY, VPZ, and CS, which may be thought of as a section of a corner of the box. In the image this corner is viewed along the direction of tilt (see Figure 7-2). This tetrahedron has three right angles at its **CS** vertex. Since angle VPY-CS-VPX is a right angle (in the horizontal reference plane), then **CS** is on a circle (in the same reference plane) with VPX-VPY as a diameter. Similarly, CSXY is also on a circle with VPX-VPY as a diameter, and this circle lies in the image plane. The relation of **CS** to CSXY (via revolution about THLXY) places **pp** on the line that is perpendicular to THLXY and contains the point CSXY.

Next, we show that **pp** is the intersection of the altitudes of triangle VPX-VPY-VPZ. By symmetry, if **pp** is on one of the altitudes, it is on all of them, so it will suffice to show that **pp** is on the altitude from VPZ.

This property becomes readily evident by visualizing the situation in three dimensions, and representing as vectors the differences between the points. The vector **CS-VPZ** is perpendicular to the XY plane generated by the vectors **CS-VPX** and **CS-VPY**. Thus vector **CS-VPZ** is perpendicular to the line THLXY. Also, the vector **CS-pp** is perpendicular to the image plane, and hence is perpendicular to THLXY. Since vectors **CS-VPZ** and **CS-pp** are both perpendicular to THLXY, their difference, VPZ-**pp**, is also perpendicular to THLXY. Hence **pp** is on the altitude of triangle VPX-VPY-VPZ from VPZ.

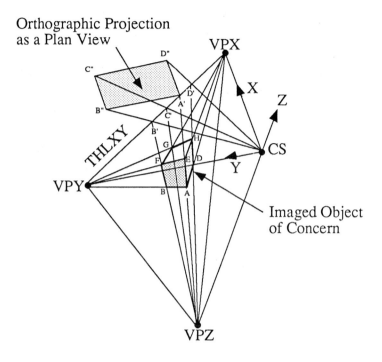

Orthographic Projection as a Plan View

Imaged Object of Concern

Figure 7-2. Illustration of Plan-view Construction showing VPX-VPY-VPZ-CS Tetrahedron

Note that the preceding paragraph is an alternative proof to the plane-geometry theorem (Perfect, 1986) that the altitudes of a triangle are concurrent. The proof achieves its simplicity by making the triangle (represented above as VPX-VPY-VPZ) the base of a tetrahedron whose apex forms a rectangular corner (represented above as **CS**).

Having justified geometrically the first three steps of our reconstruction — involving the placement of the vanishing points, CSXY, and **pp** — we now justify the steps that actually produce the orthographic views. The points A', B", C", and D" are the projections of the corners of the top of the box when the box is moved in the +Z direction (in object space) until the top of the box lies in the horizontal reference plane (XY horizon plane) containing **CS**. (Here, the positive Z direction is represented by AE,

DH, BF, and CG in the image, and raising the box is a construction equivalent to projecting the box orthographically onto the horizontal reference plane.) Now imagine the same vertical translation of the box in object space, with the visual rays from the **CS**. When the top becomes coincident with the horizon plane, with the visual rays passing through the THLXY, the points A', B', C', D' are constructed on THLXY. Likewise, to project the vertical lines FA, ED, and GB from VPZ to the THLXY will produce the same A', B', C', D' points. Then, by revolving the image plane about THLXY into the XY horizon plane (with CSXY revolved back to CS), reveals that the visual rays from CSXY through A', B', C', and D' intersect with the corresponding object-space corners of the raised box. To actually follow these steps, using object space, would produce a **1:1** orthographic projection. Since this is not practical, the concept of a model space is used.

A model space contains a scale model of the projected space of concern, obtained by projecting object space points along the visual rays, while maintaining their relative orientation (see Figure 7-3). Model space is on the opposite side of the positive image

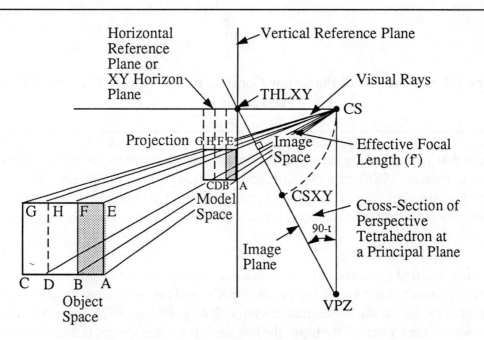

Figure 7-3. Object Space, Model Space, and Image Space

plane from the CS. Model space preserves the angular dimensions of the visual rays from the CS, allowing the construction of practical orthographic projections. In Figure 7-2, the model space for the box is selected so the corner A of the box projects orthographically to the line THLXY. The other points of the model project orthographically to the XY horizon plane to produce the plan view A'B"C"D" as illustrated in Figure 7-2. Revolving the XY horizon plane about THLXY until it coincides with the positive image plane produces the construction in Figure 7-1. Each edge of the rectangle ABCD is parallel to either CS-VPX or CS-VPY, hence the orthographic plan view of the rectangle is made by locating the box model space corner A" at A' (on THLXY) and constructing the parallel lines through point A'. The orthographic corners B", C", and D" are constructed by the projecting lines from CSXY through the points on THLXY, as described in Step 5 and Step 6 in Section 7.2.

7.4 Retrieving Other Parameters of Resection

The procedures in Step 1 through Step 3 of Section 7.2 reconstruct the principal point of the image and the revolved camera station CSXY. The effective focal length **f'** is readily determined from this construction. Also retrievable are the scaled camera-station coordinates in the object-space coordinate system generated at one corner of the solid (e.g., the corner at A). Finally, the attitude angles of the camera (azimuth **a**, tilt **t**, and swing **s**) are retrievable relative to this coordinate system.

7.4.1 Camera Station Coordinates

The scaled object-space X and Y coordinates of the camera station can be measured shown in Figure 7-1 using the plan view points A', B", C", D", and CSXY. Similarly, once the plan views of the box are brought to a common scale, the Z coordinate of the camera station can be measured using the plan views of other faces of the box, which are constructed by the same projection procedure with **THLYZ** and **THLXZ** instead of **THLXY**.

7.4.2 Effective Focal Length

To retrieve the relative focal length **f'** (with units of the image coordinates), draw a line through VPZ and **pp** and extend this line to intersect THLXY at point TXY (see Figure 7-4). Then, draw a semicircle whose diameter is VPZ-TXY. Construct a line perpendicular to VPZ-TXY at the point **pp** and label as CSRXY the intersection of this line with the semicircle. The distance from CSRXY to **pp** is clearly the effective focal length **f'**, for CSRXY is the image point obtained by revolving the camera station 90 degrees about the line VPZ-TXY. This can be checked by drawing an arc through CSXY with TXY as the radius point. The arc should pass through CSRXY.

7.4.3 Rotation Angles

The camera-attitude angles can also be determined as shown in the construction of Figure 7-4. The tilt angle **t** is defined as the angle between the image-plane normal (principal ray) and the object-space Z axis (nadir direction) in the principal plane. From the definition of CSRXY in Figure 7-4, it follows that the angle **pp-CSRXY-VPZ** is the tilt **t**, as shown.

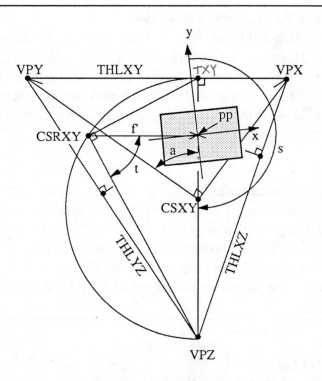

Figure 7-4. Locating the Camera Attitude Angles

The azimuth **a** is defined as the angle between the object-space X axis and the line of intersection (**THLXY**) of the image plane with the object-space XY plane. The plan view A', B", C", D', in Figure 7-1 reveals **a** to be the angle between A'-B" and A'-VPX (or between THLXY and VPX-CSXY). The later definition is equivalent to the angle denoted **a** in Figure 7-4. Expressed in object space terms, **a** can be defined as the angle between the object-space XZ plane and the plane normal to the XY plane through the THLXY, as seen in XY plan view.

The swing **s** is the easiest angle to visualize, for it represents a transformation of coordinates within the image plane and not an orientation change of the plane itself. Swing is defined as the angle between the image-space +y axis and the line of intersection between the image plane and the principal plane (generated by the principal ray and the nadir direction). By convention (Slama, 1980), **s** is the angle of clockwise rotation about the principal ray from +y to the principal plane. Swing is readily found by drawing a ray in the image-space +y direction through **pp**, measuring the angle this line makes with the ray **pp**-VPZ, and denoting this angle as **s** as in Figure 7-4.

7.5   Extension to Solids with Two Parallel Nonrectangular Faces

The methods of this chapter can be applied, in slightly modified form, to structures (such as the Pentagon in Washington, D.C., or the Flatiron building in New York City) whose walls are rectangular, but whose floors and ceilings do not form rectangles. Finding the camera parameters and plan views in these examples requires one known quantity to replace the rectangularity of the box top invoked earlier. In the example of the Pentagon (see Figure 7-5a, E-6), either one interior angle of the pentagon must be known, or a camera-attitude angle such as tilt **t** must be known.

We discuss the analyses proceeding from these two beginnings as separate cases (see Figures 7-5 and 7-6), and use for illustration a structure whose top and bottom are regular pentagons.

7.5.1 Locate CS and VPX with Known Object Angle

If a pentagon angle **p** is known (see <215 in Figure 7-5), draw the vanishing points VPZ, VP12, and VP15 by extending parallel lines until they meet. Here we are using single lines in the figure to represent sets of parallel lines. For example, line 1-2 represents a set of parallel lines from the tops and bottoms of windows, the parapet lines, and so on, which are omitted from the figures to simplify the illustrations; these lines converge at VP12. Arbitrarily assign object-space Y in the direction of the line 1-2 (so VPY = VP12).

The first objective is to find VPX, the vanishing point for lines parallel to the plane of the pentagon and perpendicular to edge 1-2. VPX must lie on the line VPY-VP15 because all sets of parallel lines in the same plane will converge to vanishing points on a common horizon line. The line VPY-VP15 is hence THLXY, and the graphical

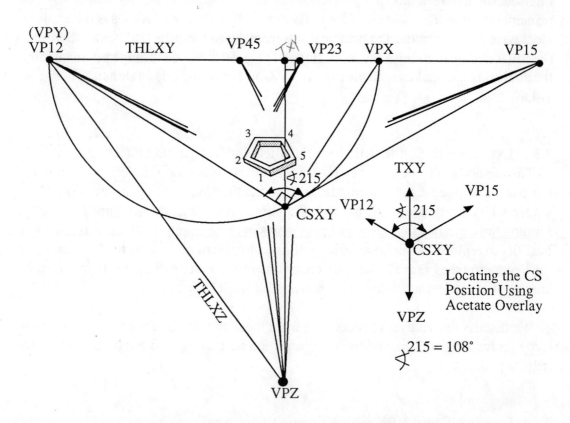

Figure 7-5. Finding Plan View of Pentagon: Pentagon Angle Known

accuracy with which it is drawn will be enhanced by drawing the other vanishing points VP45, VP23 as indicated in Figure 7-5. Again, it should be noted that these vanishing point locations are found using sets of parallel lines which are represented by a single line (for simplicity) in the illustrations of Figures 7-4 and 7-6. Now draw the perpendicular from VPZ to THLXY, and designate the intersection TXY. On a piece of clear plastic, draw two lines intersecting at angle **P**, and place the intersection point on the line VPZ-TXY. While keeping this intersection point on that line, adjust the position of the piece of plastic until the lines scribed on the plastic pass through VPY and VP15. The position of this intersection point on the line VPZ-TXY is CSXY, the camera station revolved 90 degrees about THLXY. A perpendicular to the VPY-CSXY drawn through CSXY will then intersect with THLXY at VPX (see Figure 7-

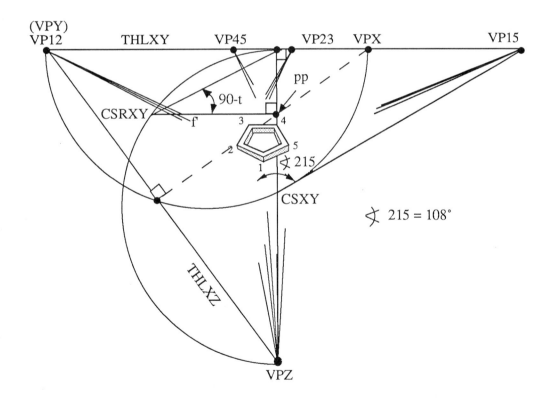

Figure 7-6. Finding Plan View of Pentagon: Tilt Angle Known

5). Once VPX is determined, the   construction of plan views proceeds analogously with the presentation in Paragraph 7.2.

7.5.2 Locate VPX and CS with Known Tilt

If the tilt angle is known (see Figure 7-6), construct VPY, VP15, VPZ, and TXY as before, and then draw the circle whose diameter is VPZ-TXY. Draw a ray from TXY making angle **t** with respect to the line VPZ-TXY; label as CSRXY the intersection of this line with the semicircle. Drawing the perpendicular to VPZ-TXY through the point CSRXY locates the principal point **pp** on VPZ-TXY as shown. A line drawn through **pp** perpendicular to THLYZ intersects with THLXY at point VPX. The plan-view reconstruction then proceeds as before. This construction from the tilt angle can be understood in the same way as the other constructions, given that CSRXY represents the camera station revolved 90o about VPZ-TXY, and **pp** is the intersection of the altitude of the triangle VPX-VPY-VPZ.

7.6 Conclusion

In this chapter, we have used specific aspects of single photograph perspective (e.g., the shapes of manmade objects) to present a review of graphical techniques commonly used in photogrammetric analysis of three-point perspective photography. Normally, these graphical techniques (elaborated in books and journals on architectural drawing) are used in transforming orthographic views into perspective views. With a few modifications, we have adopted these techniques to transform close-range perspective photography into orthographic views. The close-range perspective photograph is actually an auxiliary view of the area or object of concern. Given this auxiliary view, the photogrammetric analyst can use graphical perspective techniques (with all due concern for accuracy) to construct the orthographic plan and elevation views. An added note to those who would make use of these techniques on an irregular basis: graphical perspective techniques are an acquired skill and that means you should practice, practice, and then practice some more.

DIMENSIONAL ANALYSIS THROUGH PERSPECTIVE
Chapter Eight
Definitions

The following terms are defined for reference in this manual and relate to the Phase ONe and Phase Two procedures defined herein. The definitions have been obtained from various sources, including the Fourth Edition of the Manual of Photogrammetry and other publications listed in the references. It should be noted that lower-case letters refer to image space values and upper-case letters refer to object-space values.

AZIMUTH (**a**): The angular displacement of the principal plane of the photograph, measured clockwise to the positive Y-axis. The angle **a** is azimuth, and ranges from 0° to 90°. The co-angle is the angle at the VPY, between the Y-axis and the true horizon line.

CAMERA AXIS: A line perpendicular to the focal plane of the camera (the image plane) and passing through the perspective center. The intersection of the image plane with this line is the principal point.

CAMERA STATION (**CS**): The point in space occupied by the camera lens at the moment of exposure. This point acts as the perspective center of the photograph.

FIDUCIAL MARKS: Index marks, usually four, which are geometrically connected with the camera lens through the camera body and are usually used to define the principal point. Sometimes two widely spaced points, per format edge of full-frame images, are measured and used as temporary fiducial marks.

FORMAT: The extent of an image from edge to edge in both directions (length and width). A full format image is an uncropped print of the original negative, whether it is enlarged or not. To ensure full format for prints, requestors often ask for the inclusion of sprocket holes.

HORIZON TRACE: A straight line in the plane of the image defining the intersection of the image plane and the true horizon plane. There are three horizon traces, one for each of the three object-space coordinate planes (XY, YZ, and XZ). Each trace is drawn between pairs of vanishing points. The horizon trace of the XY plane is referred to as the true horizon line (THLXY).

IMAGE: The negative or reproduction (film or paper print) of the negative. Often the terms image and photograph (photo) are used interchangeably, even though image refers to film (see single-photograph).

IMAGE ATTITUDE: The angular orientation of the image coordinate system with respect to the object-space coordinate system. The three angles for single-image perspective are azimuth, tilt, and swing.

IMAGE COORDINATE AXES: The x-axis is parallel to the bottom edge of the full-frame format, and the y-axis is perpendicular to the x-axis. The principal point is considered the origin of this system and the z-axis is perpendicular to the xy plane and coincident with the focal length. The x and y coordinate axes are in the image plane.

IMAGE PLANE: The plane coincident with the film at the time of collection, and coincident with the photographic print or film positive at the time of mensuration.

IMAGE SPACE: Three-dimensional space inside the camera, generally described by the interior orientation elements and the image plane.

INTERIOR ORIENTATION ELEMENTS: The principal-point coordinates (xpp, ypp), the focal length (f), and the calibration distortion-correction parameters. In single-image perspective, it is rare to have the calibration parametric values. Often the center of format for full-frame imagery is used as the best approximation for the principal point (pp).

OBJECT-SPACE COORDINATE AXES: Three mutually orthogonal axes defining a right-handed rectangular object-space coordinate system. The system may be relative (defined by the user) or absolute (as in some state-plane coordinate systems). Each pair of axes defines one object-space coordinate plane.

ORIENTATION MATRIX: The 3 x 3 orthogonal matrix, designated by the symbol [R], that gives the direction-cosine values of the principal ray in the object-space system. The elements of [R] are the nine direction-cosine values.

PERSPECTIVE CENTER: The point in space through which the bundle of visual rays from object space pass to enter image space. The basic geometrical assumption in photogrammetry is that the object point, the perspective center, and the image point, all lie on the same straight line.

PHOTOGRAPH (PHOTO): A paper print reproduction of a negative. The photo is usually an enlargement of the negative.

PRINCIPAL LINE: The trace of the principal plane on the image plane. The principal line contains the principal point, the point THxy, and the nadir point (Z vanishing point).

PRINCIPAL LINE RIGHT TRIANGLE (PLRT): There are three PLRT's, the most important of which is the triangle formed in the principal plane by the points THxy, CS, and VPZ.

PRINCIPAL PLANE: The plane perpendicular to the image plane containing the principal line and the camera station. This plane also contains the geometric description of the tilt angle.

PRINCIPAL POINT (pp): The intersection of the principal ray and the image plane. This is usually the origin of the image-space coordinate system.

PRINCIPAL RAY: The ray from the camera station that intersects the image plane at right angles.

SINGLE-PHOTO: A photographic paper print of a negative. The term single-photo and single-image are sometimes used interchangeably even though single-image is often used to refer to a film image.

SWING (s): The swing angle is the angle measured clockwise from the positive direction of the principal plane (above the x-axis) to the negative y-axis. The angle s is swing, and ranges from 90° to 270°.

TERRESTRIAL IMAGERY: Any imagery collection, from any platform or source, that has limited or inadequate parametric ephemeral data for close-range photogrammetric analysis, and is not originated electronically. This definition has been specifically enlarged to include numerous collections that have been named after the collection platform rather than the photogrammetry used in dimensional analysis. Older definitions include the statement that the collection system is actually located close to the surface of the Earth (and did not include any other celestial body).

TILT (**t**): The tilt angle is a rotation about a line parallel to the XY true horizon line and through the camera station. The angle **t** is tilt, it is always positive, and ranges from 0° (pointing at VPZ/nadir) to 180° (pointing at VPZ/zenith). A **t** angle of 90° is always pointed directly at the XY true horizon line and the **pp**.

TRUE HORIZON LINE (**THL**): One of three perspective reference lines representing the intersection of the three orthogonal geometry planes and the image plane. Normally, the three geometry reference planes correspond to the major planes of the standard orthogonal coordinate system - XY plane (horizontal plane - THLXY), the XZ plane (vertical Plane - THLXZ), and the YZ plane (vertical plane - THLYZ). The position of these THLs depends upon the perspective geometry involved. In one-point perspective there are only two THLs - THLXY (or THLH), and THLXZ (**a** = 90°) or THLYZ (**a** = 0°), also identified as THLV. Two- and three-point perspective have all three THLs.

VANISHING LINE: (1) Any line in image space converging to a specific point in image space. (2) The horizon line in the photograph upon which lie all intersection points of all lines parallel to a specific plane in object space. This vanishing line is located by intersecting the image plane with a plane through the perspective center parallel to the specified plane in the object space (XY, XZ, or YZ). In particular, the vanishing line of the horizontal (XY) coordinate plane, also known as the true horizon line, passes through vanishing points for X and Y (right and left, respectively). The same applies to vanishing lines between X and Z, and Y and Z vanishing points.

VANISHING POINT: The point on the image toward which all common parallel lines in image space converge. These parallel lines may be horizontal, vertical, diagonal, or at any angle. Some vanishing points generated and used in two-point perspective may not be generated in three-point perspective, because of the assumptions made of the geometry.

DIMENSIONAL ANALYSIS THROUGH PERSPECTIVE
Chapter Nine
Analytical Perspective

9.1 Introduction

There are two reasons for completing single-image photogrammetry using analytical rather than graphical procedures. One reason is that measurements are collected within the image format, thus do not require large drafting areas. The more important reason is accuracy. The purpose of this section is to provide the photogrammetric analyst with guidelines and worksheets to support the three-dimensional perspective analysis on a single two-dimensional image. The intent is to aid the analyst in obtaining the best possible accuracy.

Much useful information can be extracted from the perspective geometry of imagery containing man-made objects. It is often feasible to obtain from these close-range perspective images the dimensions of many engineered structures, and also of any other object with a recognizable shape. Typically, there is very little or no information available about the exterior orientation relative to the object imaged. Additionally, many of the images may have no known interior orientation, often being cropped enlargements of negatives obtained with cameras of unknown focal length or origin. Under these conditions, it is difficult, if not impossible, to extract reliable mensural data from the images by the conventional multiple-frame techniques established for cartography. An approach must be used that takes advantage of the geometric properties of the objects in a single image (Gracie, et al, 1967). Single-image perspective can be used in this regard because man-made objects, particularly engineered structures, have many common, useful geometric elements such as parallel, orthogonal, and equally segmented lines and planes. These elements are also used in multiple-image techniques, but here the emphasis is on recognizing planes and inferring their positions and attitudes in three dimensions. By exploiting the properties of a single image, it becomes possible to establish the image parameters. This in turn enables object dimensions and other three-dimensional data to be determined. It is standard practice to divide photogrammetric parametric data into interior, exterior, and object categories. The interior parameters (comprising information about the image and camera independent of object-space values) include the effective focal length, principal point coordinates, image coordinates of points on the object, coordinates of fiducial points, the refraction of the light rays by the lens elements, and distortions caused by the film and its processing. The exterior parameters (comprising information about the acquisition geometry in object space) include the object-space coordinates of the collection system, the rotation angles of the camera in some relative or absolute coordinates, and possibly the refraction due to the various optical elements traversed by the light rays. The object data consists of a priori constraints on geometry to be used in the solution. The procedures discussed in this manual use only a limited number of the possible parameters—conveniently divided into Phase One and Phase Two parameters. Phase One parameters are required to obtain Phase Two parameters, and these in turn are required to obtain object-space coordinates and/or dimensions of an imaged object. The phases are described in the following paragraphs.

9.2   Ranking by Phase

The procedures in this manual are ranked according to the parameters required to determine the interior, exterior, and object values.

9.2.1 Phase One Parameters

Phase One parameters are used in the initial solution of the perspective photograph. The initial solution involves the geometry of the perspective. The parameters are:

- The image coordinates of the major vanishing points
     (**VPX**x, **VPX**y, **VPY**x, **VPY**y, **VPZ**x, and **VPZ**y)

- The interior orientation values
     (xo, yo, and ef or **f'**)

- The rotation matrix of the image
     ([**R**])

- The rotation angles
     (**a, t, s**)

The values of these parameters can be provided or calculated. The calculated values are relative to the imagery at the time of the exposure, whereas the values provided are more likely to be nominal or estimated values of the parameters. All parametric values—whether input or calculated—should have an accompanying accuracy statement, for a dimension without an accuracy statement is worthless.

9.2.2 Phase Two Parameters

Phase Two parameters are used in the final solution of the perspective imagery. In the final solution parameters from Phase One solutions, along with known object-space parametric data, are used to determine other object-space parametric values:

- The object-space coordinates of the camera station
     (Xc, Yc, Zc)

- The object-space coordinates of a point
     (X_j, Y_j, Z_j)

- The object-space distances between points
     (D_{jk})

- Assorted parametric values used in the calculations (such as geometric forms, including circles and squares)

Phase Two parameters can be known values obtained through various resources, they can be values determined using Phase One parametric values, or they can be a mixture of known and computed values.

9.3   The Analytical Presentation in this Manual - Worked Examples

The procedures for 25 analytical example problem sets are presented in the following chapters. These worked examples demonstrate the various analytical application methods presented in Chapters Ten through Twenty-one. The one-point perspective problem will be the hangar and cart example presented in Chapter Five. The two- and three-point perspective problems will use a single example with multiple problem sets. That example is a scene of a building (warehouse), a loading dock with various objects (crates) as objects of interest, and a flatbed trailer with a large thin rectangular object (sign). The analytical procedures we have selected are given as the required steps to obtain a solution, without listing the algorithm development. Some of the algorithms represent alternate methods for obtaining the same parameter. Several methods are presented because one could be the only method that will work with the imagery (geometry) of your problem set.

Many of the math models in this manual are adapted from the references listed in the Appendix; in particular, the three-point perspective procedures are adapted from Gracie, et al (1967). Likewise, not all possible situations are described in this manual, but enough information is given so a photogrammetric analyst can generalize the techniques to similar types of perspective imagery.

9.4   Conclusion

The analytical analysis of perspective has the distinct advantages of measurements being collected within the area of the image format, and the results having greater accuracy than the graphical procedures. The analytical results have two phases, and Phase One parametric values must be determined prior to Phase Two parametric values. The examples in the Appendix use all of the math models in one form or another.

DIMENSIONAL ANALYSIS THROUGH PERSPECTIVE

Chapter Ten
Axis Conventions and Rotations

The conventions in this manual are not unique. They are those used by the authors over many years, and it is intended that they should become convenient and familiar to the readers. There may be some cumbersome transitions, but overall the conventions are easy to learn and they work, so we have kept the status quo. If a photogrammetric analyst wants to use different conventions, then the conversion from one system to the other rests with that individual, and it is suggested that the conversion must be complete.

10.1 Axes

The relationship between the image- and object-space coordinate systems is shown in Figure 10-1. Since the perspective imagery presented in this manual involves relative rather than absolute positions and dimensions, any convenient right-handed object-space coordinate system can be used. The object-space coordinate system shown in Figure 10-1 is the system selected for all procedures in this manual. The positive Z-axis is directed vertically upward and perpendicular to the horizontal plane (XY plane). The positive X-axis is directed toward the right and away from the camera station, and the positive Y-axis is directed toward the left and away from the camera station.

The origin of the object-space coordinate system is defined to be in the area covered by the imagery. Thus the two camera-station coordinates, Xc and Yc, are always negative. The third camera-station coordinate Zc can be either positive (Figure 10-1) or negative, depending on whether the camera station is above or below the horizontal plane. Of course, the selection of the origin is arbitrary; however, for convenience, the photogrammetric analyst should position the origin anywhere that will leave all the calculated values with a positive sign. If only dimensions (not coordinates) are known in object space, it is quite possible to assign the coordinate values for the camera station and thus define the origin in a more useful manner. As you work through examples it will be apparent why defining the coordinate system in this manner is desirable.

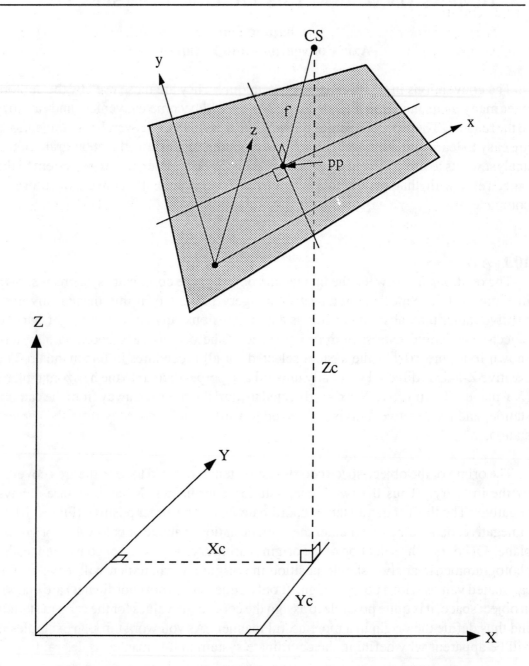

Figure 10-1. Image- and Object-Space Coordinate Systems

10.2 Rotations

There are three rotation angles used in determining the attitude of the camera for single-image perspective as presented in this manual. These angles are azimuth, tilt, and swing. The azimuth angle (**a**) of the image always lies between 0° and 90° under this particular axis convention. Angle **a** is the clockwise rotation from the object-space Y-axis to the principal plane, a rotation about the object-space Z axis. The tilt angle (**t**) of the image lies between 0° and 180°, with a tilt angle of 90° occurring when the principal ray is pointed directly at the true horizon trace. Angle **t** is a positive rotation about an axis parallel to the true horizon line (XY horizon trace), with nadir being 0° tilt. The swing angle (**s**) is measured from the positive side of the principal line to the negative image y axis. Angle **s** always lies between 90° and 270°. When the horizontal edge of the original full-format frame is parallel to the XY horizon trace, the swing angle is 180°. The swing angle is *a* clockwise rotation about the principal ray of the camera.

10.3 Conclusion

The conventions used in defining the image- and object-space coordinates are conventional right-handed coordinate systems. The rotation angles follow the same construction. Procedures for determining these parametric values are described in the following chapters, and examples are worked in the Appendix.

DIMENSIONAL ANALYSIS THROUGH PERSPECTIVE

Chapter Eleven
Phase One Parametric Procedures

11.1 Introduction

Presented in the following Chapters (Twelve through Sixteen) are 21 analytical methods used in the first phase of the photogrammetric analysis solution. These methods involve finding specific parametric values. The general derivations for determining these parametric values are presented in this Chapter. In holding to our statement in the Preface, and section 1.3, we have deliberately not given a detailed treatise on the derivations here or throughout this manual. The analytical methods involve finding the following parametric values:

1. The image coordinates of the vanishing points of the object-space coordinate axes (x_1, y_1, x_2, y_2, and x_3, y_3).

2. The image coordinates of diagonal vanishing points and measuring points (x_n, y_n — where n is any number larger than 3).

3. The interior orientation elements of the imagery (**xo**, **yo** and **f'**), and the image camera station coordinates(**CSx** and **CSy**).

4. The elements of the imagery orientation matrix, [**R**].

5. The azimuth (**a**), tilt (**t**), and swing (**s**) of the imagery in the imagery-to-ground orientation.

11.2 Projective Equations

The general equations used in the formulation of single-photo perspective procedures are the projective equations

$$\begin{bmatrix} x_j - xo \\ y_j - yo \\ 0 - f'c \end{bmatrix} = Sc\,[R] \begin{bmatrix} X_j - Xc \\ Y_j - Yc \\ Z_j - Zc \end{bmatrix} \qquad (Eq\ 11\text{-}0)$$

where the left side of the matrix equation is the image-space vector, and on the right side are the undetermined scaler, the rotation matrix, and the object-space vector.

    The projective equations define the projected visual rays between the object-space coordinate system and the image coordinate system. The scale factor (conversion factor), Sc, may be defined as the ratio of an image length to the corresponding object-space length; e.g., effective focal length to range (Sc = f'/R). This ratio, along with the orientation matrix, enables the conversion of object-space values to image-space values or image-space values to object-space values. The orientation matrix, [R], defines the direction of the principal ray relative to the vertical (down) direction. Matrix [R], a 3x3 matrix, contains the nine direction cosine values for the principal ray. A matrix is developed for each rotation angle (a, t, and s). The rotation matrices are then multiplied in the order of s, t, and a, to obtain the final rotation matrix [R]. Explicitly,

$$(Eq\ 11\text{-}1)$$

$$[R] = \begin{bmatrix} -\cos(s) & -\sin(s) & 0 \\ \sin(s) & -\cos(s) & 0 \\ 0 & 0 & 1 \end{bmatrix} \begin{bmatrix} 1 & 0 & 0 \\ 0 & \cos(t) & \sin(t) \\ 0 & -\sin(t) & \cos(t) \end{bmatrix} \begin{bmatrix} \cos(a) & -\sin(a) & 0 \\ \sin(a) & \cos(a) & 0 \\ 0 & 0 & 1 \end{bmatrix}$$

The elements determined for the rotation matrix [R] will equal the values of any other close-range rotation sets; e.g., omega, phi, and kappa; or pitch, roll, and yaw; and so on, used to define the orientation of the imagery.

    It is possible to use a number of math models to describe the projection equations. There are times when it was deemed necessary to use the equations of the graphical lines to provide the user a check on a graphical solution that may have already been completed. Therefore, the math models used for this manual are a combination of the equations given above and the equations that describe the layout of the possible graphical solutions.

11.3 Vanishing Points

It is useful to trace here some of the mathematical development leading to vanishing points. If a camera looks at a scene in object space, the familiar collinearity equations connect the object-space coordinate (X, Y, Z) of a point to the image-space coordinates (x, y).

From the collinearity equations, we can show that parallel lines in object space map to lines that are either parallel or all converge to a single point (vanishing point) in image space. To prove this theorem, we first write a parametric equation for a line in object space through a point (Xo, Yo, Zo) that is parallel to a unit vector (QX, QY, QZ):

$$\begin{bmatrix} X \\ Y \\ Z \end{bmatrix} = \begin{bmatrix} Xo \\ Yo \\ Zo \end{bmatrix} + T \begin{bmatrix} QX \\ QY \\ QZ \end{bmatrix} \qquad \text{(Eq 11-2)}$$

where T is the parameter whose variation generates the points on the line. Parallel lines are lines with different (Xo, Yo, Zo) but the same (QX, QY, QZ). An image of a point on a line is represented by substituting expressions for X, Y, and Z from Eq. 11-2 into Eq. 11-0.

Far from the camera station, in the limit $T \longrightarrow \infty$, Eq. 11-0 becomes

$$\frac{x - xo}{-f'} = \frac{r_{11}QX + r_{12}QY + r_{13}QZ}{r_{31}QX + r_{32}QY + r_{33}QZ} \qquad \text{(Eq 11-3a)}$$

$$\frac{y - yo}{-f'} = \frac{r_{21}QX + r_{22}QY + r_{23}QZ}{r_{31}QX + r_{32}QY + r_{33}QZ} \qquad \text{(Eq 11-3b)}$$

Note that Eq. 11-3 is independent of (Xo, Yo, Zo), so the point (x, y) represented by Eq. 11-3 is the intersection point of all lines parallel to (QX, QY, QZ). This vanishing point, of course, may be at infinity if the denominator is zero.

The vanishing points for lines parallel to the object-space cartesian coordinate axes are given by substituting into Eq. 11-3 the respective unit vectors $(1,0,0)$, $(0,1,0)$, and $(0,0,1)$:

$$\frac{VPXx - xo}{-f'} = \frac{r_{11}}{r_{31}} \quad ; \qquad \frac{VPXy - yo}{-f'} = \frac{r_{21}}{r_{31}} \quad ; \qquad \text{(Eq 11-4a)}$$

$$\frac{VPYx - xo}{-f'} = \frac{r_{12}}{r_{32}} \quad ; \qquad \frac{VPYy - yo}{-f'} = \frac{r_{22}}{r_{32}} \quad ; \qquad \text{(Eq 11-4b)}$$

$$\frac{VPZx - xo}{-f'} = \frac{r_{13}}{r_{33}} \quad ; \qquad \frac{VPZy - yo}{-f'} = \frac{r_{23}}{r_{33}} \quad . \qquad \text{(Eq 11-4c)}$$

If r_{31}, r_{32}, and r_{33} are all nonzero, then there is three-point perspective, for in that case VPX, VPY, and VPZ all have (finite) vanishing points. If one of the quantities r_{3k} is zero (where k = 1,3), two finite vanishing points remain, and there is two-point perspective. In this case, the k-th coordinate axis is parallel to the image plane. If two of the quantities r_{3k} are zero, then two vanishing points are at infinity, and only one vanishing point is finite. In this case, called one-point perspective, only one of the coordinate axes is nonparallel to the image plane. Note that, for one-point perspective, the principal point (xo, yo) is the finite vanishing point. [For example, let $r_{32} = r_{33} = 0$; then $r_{31} = 1$ because R is an orthogonal matrix. But orthogonality of R also implies $r_{11} = r_{31} = 0$, hence from Eq. 11-4 it follows that VPXx = xo and VPXy = yo.]

11.4 Point Selection

During the Phase One analysis the photogrammetric analyst will select a large number of points: points on vanishing lines, points to define geometry and points to define the object of interest. Just as in a graphical solution there is a need to know the basic principles of drafting (sharp pencils and all that), there is a need in analytical procedures to understand point selection (**PS**). The major items to remember are:

PS-1. Points on the format edge for fiducial measuring must not be too close to the corners of the frame format.

PS-2. Points on vanishing lines must not be too close together: short lines have large errors.

PS-3. Points on vanishing lines that are on the edge of something must lie on the same side of the contrast.

PS-4. Systematically approach all intersection points from the same direction — light to dark or dark to light — to be consistent.

PS-5. If there is good resolution do not multiple-point a line — just use two points.

PS-6. If the resolution is not good it may be wise to use the multiple-point technique and use the average of the pointings for the one point in your calculations.

PS-7. Always identify your points using an enlarged print or overlays or an enlarged print with overlays.

PS-8. There is nothing wrong with making one point serve several purposes, but use a method to keep track of the purposes. Multiple ID numbers is one method.

PS-9. Prior to point selection, identify the methods you would use in the solution. When you select points you should keep a record of which points are to be used with which method.

PS-10. Never trust any aspect of point selection to memory; you have more important work for your brain, so keep meticulous notes.

11.5 Conclusion

Determining Phase One parameters is a necessary first step in any photogrammetric analysis of perspective. The parameters provide image-space designations (e.g., vanishing points, principal point, and so on) and camera angle determinations that will be needed in further inferences about object-space coordinates and dimensions. The accuracy of model-space reconstructions, therefore, depends critically on the accuracy of the determination of Phase One parameters, and this accuracy in turn depends on the care with which image points are selected (identified) to determine the image values. It is hoped that the present chapter has provided some hints that will facilitate the accurate determination of Phase One parameters. More details about determining particular Phase One parameters are presented in the next five chapters.

DIMENSIONAL ANALYSIS THROUGH PERSPECTIVE

Chapter Twelve
Location of the Principal Point

The principal point location is usually defined as the perpendicular intersection of the principal ray and the image plane. The principal point can also be best approximated as the center-of-format for full frame format imagery. This approximation is preferred in single-image perspective solutions unless the principal point can be obtained using perspective geometry solutions. Of the many methods available to determine the principal point, only four are presented in this manual. The first and second methods involve locating the center-of-format, and may be used for one-, two-, or three-point perspective solutions. The third method uses a diagonal of known angle, which provides the camera station and the principal point coordinates for two-point solutions and only the camera station coordinates in a three-point solution. The fourth method uses the three vanishing points to determine the location of the principal point. In each method the origin of the measuring system is initially defined to the left and below the imagery being used, with axes oriented so that all of the image coordinates are positive.

12.1 Center-of-Format Method - Known Fiducials

The equations in this subsection apply when four fiducial points (one located on each frame edge) define the center-of-format and possibly the principal point. In these equations, the first fiducial is at the top of the frame and is numbered 1 (xf_1 and yf_1). The remaining fiducials are numbered clockwise; e.g., 2, 3, and 4. If the fiducials are not in the corners and not on the frame edges, then the fiducial numbered 1 is taken to be the upper right corner intersection point, and the remaining fiducials (corners) are numbered clockwise. To determine the best estimate for the principal point (xo, yo: **xpp, ypp**), compute the intersection of lines 13 and 24 using the following procedure.

Center-of-Format (E-7)

1. $dx_{13} = xf_1 - xf_3$ = ...
2. $dy_{13} = yf_1 - yf_3$ = ...
3. $dx_{24} = xf_2 - xf_4$ = ...
4. $dy_{24} = yf_2 - yf_4$ = ...
5. $w_1 = (xf_1 * yf_3) - (yf_1 * xf_3)$ = ...

6. $w_2 = (xf_2 * yf_4) - (yf_2 * xf_4)$ = ...
7. $w_3 = (w_1 * dx_{24}) - (w_2 * dx_{13})$ = ...
8. $w_4 = (w_1 * dy_{24}) - (w_2 * dy_{13})$ = ...
9. $w_5 = (dx_{13} * dy_{24}) - (dx_{24} * dy_{13})$ = ...
10. $xo = w_3 / w_5$ = ...
11. $yo = w_4 / w_5$ = ...

It should be noted that this method determines the intersection of the two lines defined by the frame fiducials (or corners), and not necessarily the principal point. This intersection point becomes the best fit approximation for the **pp**, unless the calibrated offset values (deltas and rotation angle) are known. If the calibrated values are known, the correct adjustments should be made to the computed values (Slama, 1980).

12.2 Center-of-Format Method - Intersection of Two Lines

When there are no fiducials, and the image coordinates of the corners of the frame are required to find the center-of-format, compute the coordinates of the frame corners. To find the four corners of the frame (fiducial corners) and the intersection of the diagonals, first measure two arbitrarily selected points along each side (edge) of the frame. This will provide you with eight measured points. Identify the points 1 and 2 on each line, with point 2 being closest to the corner. Do this for each pair of lines used to determine a corner. The lines are identified clockwise as A and B. Now find the intersection of pairs of lines, each line passing through a pair of points, to determine the corners. The analyst should use a data recording form to record the measured values. Compute the values in the order shown, once for each corner.

Compute the Corner Values (E-8) √

	Corner equations		Corners C-1 C-2 C-3 C-4
1.	$dx_1 = xA_1 - xA_2$	=	....
2.	$dy_1 = yA_1 - yA_2$	=	....
3.	$dx_2 = xB_2 - xB_1$	=	....
4.	$dy_2 = yB_2 - yB_1$	=	....
5.	$w_1 = (xA_1 * yA_2) - (yA_1 * xA_2)$	=	....
6.	$w_2 = (xB_2 * yB_1) - (yB_2 * xB_1)$	=	....

7. $w_3 = (w_1{}^{\text{v}}dx_2)-(w_2{}^{\text{k}}dx_1)$ = ...

8. $w_4 = (w_1{}^*dy_2)-(w_2{}^*dy_1)$ = ...

9. $w_5 = (dx_1{}^*dy_2)-(dx_2{}^*dy_1)$ = ...

10. $xc_i = w_3 / w_5$ =   __ __ __ __

11. $yc_i = w_4 / w_5$ =   __ __ __ __

When the four corner fiducials have been determined, compute the intersection of the opposite corners (consider them fiducial corners) as described in paragraph 12.1 to determine the coordinates of the center-of-format, and the best approximation of the principal point.

12.3 Diagonal Methods - Horizontal and Vertical

12.3.1 Introduction

This method is for two-point and three-point solutions. Horizontal diagonals are used in three-point solutions when the VPZ cannot be located because of a lack of vanishing lines. The two-point solution actually produces all of the Phase One parametric values (see Chapter Seven), while the three-point solution produces a limited number of parametric values, including the x coordinate of the principal point. The rationale for this is simple. In a two-point solution the tilt angle is 90° and the principal point is on the XY horizon trace. The three-point solution has a tilt angle other than 90° and the principal point is on the principal line, which perpendicularly intersects the XY horizon trace and passes through the camera station. Both solutions use the fact that the true value of any angle can be generated at the camera station and the line of the angle will pass through the vanishing point of the diagonal located on a horizon trace (see Figure 12-1). Two-point perspective has the added capability that parallel diagonals in vertical planes will converge to a vanishing point (**VPD**) on a perpendicular line through the respective vanishing point (**VPX** or **VPY**).

The co-angle is generated from the perpendicular line towards the other vanishing point. The point at which the vertical angle line intersects the XY horizon trace is a measuring point (**MP**) which is the camera station rotated to the XY horizon trace (see Figure 12-2).

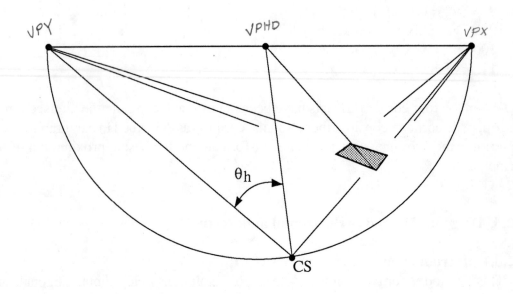

Figure 12-1. Vanishing Point for Horizontal Diagonal

12.3.2   Diagonal of Known Angle: Horizontal Plane

Several assumptions are made for this solution. First, the tilt angle is 90° for two-point procedures. Second, the orientation of the measuring system is such that the x-axis is parallel to or coincident with the **THL**. If the measured data does not conform to this assumption, then the data must be processed through a transformation routine (see Chapter Twenty-One). Points for the solution are numbered as shown in Figure 12-1. The x and y coordinates (**CSx, CSy**) of the camera station are computed. The solution is based on a unique method of calculating camera-station coordinates, given that the angle of the horizontal diagonal is known and that the horizontal diagonal vanishing point (**VPHD**) can be located. Let **VPX**, **VPY**, and **VPHD** be points 1, 2, and 3 respectively. This method will work for either two-point or three-point perspective.

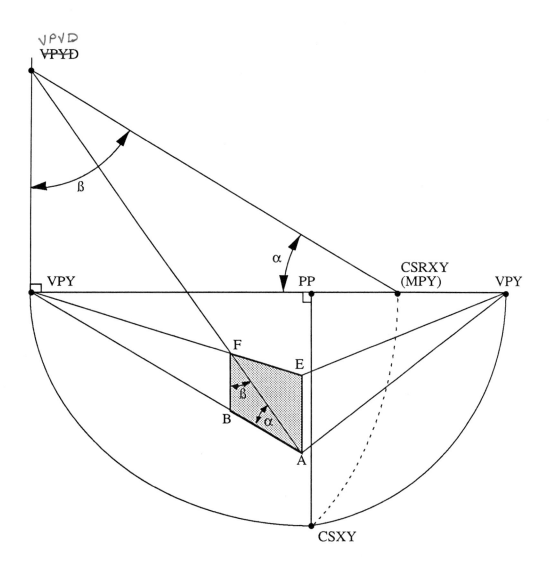

Figure 12-2. Vanishing Point for Vertical Diagonal

Principal Point using DKA-Horizontal Plane Method ✓
(see Example E-9, in the Appendix)

Known angle = **ka**, x4 = **xpp**, and x4,y4 = **CSx,CSy**

1. $w_1 = x_1 - x_2$ = ...
2. $w_2 = x_3 - x_2$ = ...
3. $w_3 = (\tan(\mathbf{ka}))^2$ = ...
4. $w_4 = w_1 * w_2^2$ = ...
5. $w_5 = w_3 * (w_1 - w_2)^2 + w_2^2$ = ...
6. $x_4 = (w_4 / w_5) + x_2$ = ...
7. $y_4 = y_2 - ((x_4 - x_2) * (x_1 - x_4))^{1/2}$ = ...

12.3.3 Diagonal of Known Angle: Vertical Plane

This is a two-point perspective solution and will not work for one- or three-point perspective. The three- point solution will use the horizontal diagonal method on one of the other horizon traces. It is assumed that the origin and orientation of the mensuration coordinate system are such that the measured coordinates are positive, and the x-axis is parallel or coincident to the **THL**.

Vanishing Point Diagonal - Vertical at VPY
Two-Point Solution (E-10) ✓

Line A is perpendicular to the horizon line. Point A_1 is **VPY**. Point A_2 x value is the same as **VPYx**, and the point A_2 y value is some arbitrary value above the **VPY**. Line E is the diagonal line with point E_2 closest to VPDV (see Figure 12-2). The following quantities are computed in the order given.

known angle = **ka**

1. $dy_1 = yA_1 - yA_2$ = ...
2. $dx_2 = xE_2 - xE_1$ = ...
3. $dy_2 = yE_2 - yE_1$ = ...
4. $w_1 = (xA_1 * yA_2) - (yA_1 * xA_2)$ = ...

5. $w_2 = (xE_2 * yE_1) - (yE_2 * xE_1)$ = ...

6. $w_3 = (w_1 * dx_2)$ = ...

7. $w_4 = (w_1 * dy_2) - (w_2 * dy_1)$ = ...

8. $w_5 = -(dx_2 * dy_1)$ = ...

9. $VPDx = w_3 / w_5$ = ...

10. $VPDy = w_4 / w_5$ = ...

11. $w_6 = Tan(ka)$ = ...

12. $MPYx = \underset{\llcorner neg}{-VPYx} + ((VPDy-VPYy)*w_6)$ = ...

Note: Points VPY (P2), semicircle center (P0), and camera station (P4) form an isosceles triangle, where the base is the calculated distance between VPY and MPY (see Figure 12-2). The sides of the isosceles triangle are the radius of the arc through the camera station. The angle at VPY can now be determined.

13. $Cos(P2) = (MPYx-VPYx)/|VPXx-VPYx|$ = ...

$Sin(P2) = (1.-(Cos(P2)^2)^{1/2}$ = ...

14. $CSx = VPYx + (MPYx-VPYx)*cos(P2)$ = ...

15. $CSy = VPYy - (MPYx-VPYx)*sin(P2)$ = ...

16. $xo = CSx$ = _____

17. $yo = VPYy$ (or y value of **THL**) = _____

Vanishing Point Diagonal - Vertical at VPX
Two-Point Solution (E-11)

This method, based on knowing a vertical diagonal angle, is the procedure used in E-10. There is a sign difference in equation 14 that <u>must</u> be noted. Chapter Twelve and Figure 12-2 contain additional information.

Line A is the perpendicular to the **THL** at VPX. Point A_2 on line A is the VPX. Point A_1 x value is same as VPX and A_1 y value is some arbitrary value above the VPX. Line G is the diagonal line with point G_1 closest to VPD. The following quantities are computed in the order given.

1. $dy_1 = yA_1 - yA_2$ = ...

2. $dx_2 = xG_2 - xG_1$ = ...

3. $dy_2 = yG_2 - yG_1$ = ...

4. $w_1 = (xA_1 * yA_2) - (yA_1 * xA_2)$ = ...

5. $w_2 = (xG_2 * yG_1) - (yG_2 * xG_1)$ = ...

6. $w_3 = (w_1 * dx_2)$ = ...

7. $w_4 = (w_1 * dy_2) - (w_2 * dy_1)$ = ...

8. $w_5 = -(dx_2 * dy_1)$ = ...

9. $VPDx = w_3 / w_5 = VPXx$ = ...

10. $VPDy = w_4 / w_5$ = ...

11. $w_6 = Tan(\mathbf{ka})$ = ...

12. $MPYx = VPXx - (VPDy-VPXy)*w_6$ = ...

Note: Points VPX (P1), semicircle origin (P0), and camera station (P4) form an isosceles triangle, where the base is the calculated distance between VPX and MPX (see Figure 12-2). The sides of the isosceles triangle are the radius of the arc through the camera station. The angle at VPX can now be determined.

13. $Cos(P1) = (VPXx-MPXx)/|VPXx-VPYx|$ = ...

       $Sin(P1) = (1.-(Cos(P1)^2)^{1/2}$ = ...

14. $CSx = VPXx-(VPXx-MPXx)*Cos(P1)$ = ...

15. $CSy = VPYy-(VPXx-MPXx)*Sin(P1))$ = ...

16. $xo = CSx$ =   _____

17. $yo = VPYy$ (or y value of **THL**)   =   _____

12.4 Orthocenter Method

The three major vanishing points form a triangle. There is a certain technique to finding the intersection of the perpendiculars to the vertices of the triangle. That intersection in the perspective triangle is the principal point. When all three vanishing points are known, compute the values below in the order given to determine the coordinates of the principal point.

Principal Point Using Orthocenter Method (E-12)

VPX: x_1, y_1

VPY: x_2, y_2

VPZ: x_3, y_3

1. $dx_1x_3 = x_1 - x_3$ = ...
2. $dx_2x_3 = x_2 - x_3$ = ...
3. $dy_1y_3 = y_1 - y_3$ = ...
4. $dy_2y_3 = y_2 - y_3$ = ...
5. $w_1 = (y_1 * dy_2y_3) + (x_1 * dx_2x_3)$ = ...
6. $w_2 = (y_2 * dy_1y_3) + (x_2 * dx_1x_3)$ = ...
7. $w_3 = (w_1 * dy_1y_3) - (w_2 * dy_2y_3)$ = ...
8. $w_4 = (w_1 * dx_1x_3) - (w_2 * dx_2x_3)$ = ...
9. $w_5 = (dy_1y_3 * dx_2x_3) - (dy_2y_3 * dx_1x_3)$ = ...
10. $xo = w_3 / w_5$ =   _____
11. $yo = -w_4 / w_5$ =   _____

12.5 Conclusion

The methods presented in this chapter to locate the principal point vary from the best approximation to the best geometric position. The center-of-format method is probably the least accurate because it depends upon the manufacturer's assembly line placing the die-cut format plate in the correct position. The diagonal and orthocenter methods are the most accurate when combined and completed independently. The principal point is an image position that is never measured; the coordinates of the principal point are always computed. The magnitude of error in the analytical methods will be determined by the errors in the measured data. This is the reason why using the diagonal and orthocenter methods in combination is the most accurate — the photo-grammetric analyst can check the results and adjust accordingly.

DIMENSIONAL ANALYSIS THROUGH PERSPECTIVE
Chapter Thirteen
Locating the Vanishing Points

13.1 Locating the Vanishing Points
The methods described here are primarily intended for the calculation of the vanishing points applying to the three object-space coordinates axes illustrated in Figure 13-1. It is possible to use the first two methods to locate other vanishing points. It is also possible to use the procedures to compute one-, two-, and three-point perspective vanishing points, but extreme care must be taken to use extremely large or small numbers in the calculations where required. (This applies to the solutions when using projective equations.) Special methods have notes to help the photogrammetric analyst decide the type of data to use in the solution. Four methods have been selected for presentation in this manual: the intersection of two parallel lines, the known ratios of a segmented line, and two methods using orthogonal lines.

13.2 Two Parallel Lines Method
The calculation of the intersection coordinates of two parallel lines is the primary tool for single-image perspective, and it could cause the most accuracy problems. There are many reasons why the two-line intersection can be inaccurate. The angle between the lines may be small, the lines used to determine a line may be short, or the image may be of poor resolution, and so on. Intersecting lines that are at right angles to each other offer the best opportunity for reliable results. They were used in the center of format solution (a good solution) and are used again in this procedure (not always a good solution). The lines are identified as A and B. The points on the lines are numbered 1 and 2, with the number 2 point being the closest to the intersection point. Solve for the following values in the order given.

Two Intersecting Lines (E-13)

1. $dx_1 = Ax_1 - Ax_2$ = ...
2. $dy_1 = Ay_1 - Ay_2$ = ...
3. $dx_2 = Bx_1 - Bx_2$ = ...
4. $dy_2 = By_1 - By_2$ = ...

5. $w_1 = (Ax_1 * Ay_2) - (Ay_1 * Ax_2)$ = ...
6. $w_2 = (Bx_1 * By_2) - (By_2 * Bx_1)$ = ...
7. $w_3 = (w_1 * dx_2) - (w_2 * dx_1)$ = ...
8. $w_4 = (w_1 * dy_2) - (w_2 * dy_1)$ = ...
9. $w_5 = (dx_1 * dy_2) - (dx_2 * dy_1)$ = ...
10. $VPx = w_3 / w_5$ = _____
11. $VPy = w_4 / w_5$ = _____

13.3 Segmented Line Method

To find the vanishing point of a line where three points define three separations (segments by ratio) are known, is fairly simple, but not necessarily accurate. The points leading away from the vanishing point are in descending order, 3, 2, and 1. Determine the following values in the order given.

Segmented Line (E-14)

1. $dx_1 = x_2 - x_1$ = ...
2. $dx_2 = x_3 - x_2$ = ...
3. $dy_1 = y_2 - y_1$ = ...
4. $dy_2 = y_3 - y_2$ = ...
5. $K = D_1 / D_2$ = ...

Note: If the ratio of K is known, then D_1 and D_2 do not have to be known.

6. $w_1 = dx_2 / dx_1$ = ...
7. $w_2 = dy_2 / dy_1$ = ...
8. $w_3 = (w_1 + w_2) /2.$ = ...

Note: If the line is straight and not parallel to one of the object-space axes, then w_1 should equal w_2. Errors in measuring will necessitate the average value (w_3) to be determined. When the line is parallel to one of the object-space axes use the appropriate w_1 or w_2 values; e.g., if the line is parallel to the X-axis, use $w_3 = w_1$.

11. $w_4 = K * w_3$ = ...
12. $w_5 = x_1 * w_4 - x_1$ = ...

X_3

13. $w_6 = y_1 * w_4 - y_1$ = ...

14. $VPX = w_5 / (w_4-1.)$ =   _____

15. $VPY = w_6 / (w_4-1.)$ =   _____

13.4 Orthogonal Line Method

Given a line with two points in the imagery directed towards the position of an unknown vanishing point, the problem is to determine the image coordinates of that vanishing point. The required parametric values are: x,y coordinates of another vanishing point, the principal point and effective focal length. The points on the line are numbered A_1 and A_2, with point A_2 being closest to the unknown **VP**. In this example it is assumed that the **VPY** is the unknown **VP** and the **VPX** is the known **VP**, but it just as well could have been a diagonal vanishing point or any other vanishing point that is unknown.

Orthogonal Line - Procedure No. 1 (E-15)

1. $w_1 = VPXx - xo$ = ...
2. $w_2 = VPXy - yo$ = ...
3. $w_3 = (w_1 * xo) + (w_2 * yo) - f'^2$ = ...
4. $w_4 = yA_2 - yA_1$ = ...
5. $w_5 = xA_2 - xA_1$ = ...
6. $w_6 = (xA_1 * yA_2) - (xA_2 * yA_1)$ = ...
7. $w_7 = (w_3 * w_5) + (w_6 * w_2)$ = ...
8. $w_8 = (w_3 * w_4) - (w_6 * w_1)$ = ...
9. $w_9 = (w_1 * w_5) + (w_4 * w_2)$ = ...
10. $VPYx = w_7 / w_9$ =   _____
11. $VPYy = w_8 / w_9$ =   _____

In Procedure No. 2 the origin of the object-space axes (xa, ya) and one point on each of the three object-space axes (xw,yw; xu,yu; xv,yv) to the vanishing points are imaged. Point **w** is on the imaged X-axis, point **u** is on the imaged Y-axis, and point **v** is on the imaged Z-axis. The origin of the image coordinate system is at the principal point. The following values should be computed in the order given.

Orthogonal Line - Procedure No. 2 (E-15)

1. $da = (xa^2 + ya^2 + f'^2)^{1/2}$ = ...

2. $r_{13} = -xa / da$ = ...

3. $r_{23} = -ya / da$ = ...

4. $r_{33} = -f' / da$ = ...

5. $w_1 = (r_{13}*xv)+(r_{23}*yv)-(r_{33}*f')$ = ...

6. $w_2 = (xv^2 + yv^2 + f'^2 - w_1^2)^{1/2}$ = ...
   (w_2 takes on the sign of yv - ya)

7. $w_3 = xv - w_1 * r_{13}$ = ...

8. $w_4 = yv - w_1 * r_{23}$ = ...

9. $w_5 = -f' - w_1 * r_{33}$ = ...

10. $r_{12} = w_3 / w_2$ = ...

11. $r_{22} = w_4 / w_2$ = ...

12. $r_{32} = w_5 / w_2$ = ...

13. $r_{11} = (1. - r_{12}^2 - r_{13}^2)^{1/2}$ = ...

14. $w_6 = (r_{12}*r_{22}) + (r_{13}*r_{23})$ = ...

15. $w_7 = (r_{12}*r_{32}) + (r_{13}*r_{33})$ = ...

16. $r_{21} = -w_6 / r_{11}$ = ...

17. $r_{31} = -w_7 / r_{11}$ = ...

18. $w_8 = r_{11}*xw + r_{21}*yw - r_{31}*f'$ = ...

19. $w_9 = r_{12}*xw + r_{22}*yw - r_{32}*f'$ = ...

20. $w_{10} = r_{11}*xu + r_{21}*yu - r_{31}*f'$ = ...

21. $w_{11} = r_{12}*xu + r_{22}*yu - r_{32}*f'$ = ...

22. $w_{12} = w_8 / w_9$ = ...

23. $w_{13} = w_{10} / w_{11}$ = ...

24. $w_{14} = (|w_{12} * w_{13} + 1.|)^{1/2}$ = ...
   if camera points down, then $w_{14} = -w_{14}$ = ...

25. $w_{15} = r_{11}*w_{12} + r_{12} + r_{13}*w_{14}$ = ...

26. $w_{16} = r_{21}*w_{12} + r_{22} + r_{23}*w_{14}$ = ...

27. $w_{17} = r_{31}*w_{12} + r_{32} + r_{33}*w_{14}$ = ...

28. $w_{18} = r_{11}*w_{13} + r_{12} + r_{13}*w_{14}$ = ...

29. $w_{19} = r_{21}*w_{13} + r_{22} + r_{23}*w_{14}$ = ...

30. $w_{20} = r_{31}*w_{13} + r_{32} + r_{33}*w_{14}$ = ...

31. $w_{21} = r_{12} * w_{14} - r_{13}$ = ...

32. $w_{22} = r_{22} * w_{14} - r_{23}$ = ...

33. $w_{23} = r_{32} * w_{14} - r_{33}$ = ...

34. $\text{VPXx} = -f' * w_{15} / w_{17}$ =   _____

35. $\text{VPXy} = -f' * w_{16} / w_{17}$ =   _____

36. $\text{VPYx} = -f' * w_{18} / w_{20}$ =   _____

37. $\text{VPYy} = -f' * w_{19} / w_{20}$ =   _____

38. $\text{VPZx} = -f' * w_{21} / w_{23}$ =   _____

39. $\text{VPZy} = -f' * w_{22} / w_{23}$ =   _____

13.5 Interior Parametric Values and Rotation Matrix Method

The problem is to find the three vanishing points given the principal point, focal length, and rotation matrix. This is a useful procedure given full format imagery and imagery with which to make some good assumptions. With some assumptions the photogrammetric analyst can estimate values for the three rotation angles, use the procedure in Chapter Fifteen (15.2) to determine [R], and compute the image coordinates for the three vanishing points to determine all the Phase One values to some satisfying accuracy. The method becomes an iterative procedure when one of the VPs can be determined using the vanishing point methods of Chapter Thirteen. This known VP is then used as a value check against the computations presented here. For the example given the known values as stated. Determine the following values in the order given.

pp, f', and [R] are known - full format image (E-16)

1. $w_1 = r_{11} / r_{31}$ = ...

2. $w_2 = r_{21} / r_{31}$ = ...

3. $w_3 = r_{12} / r_{32}$ = ...

4. $w_4 = r_{22} / r_{32}$ = ...

5. $w_5 = r_{13} / r_{33}$ = ...

6. $w_6 = r_{23} / r_{33}$ = ...

7. $\text{VPXx} = xo - f' * w_1$ =   _____

8. $\text{VPXy} = yo - f' * w_2$ =   _____

9. $\text{VPYx} = xo - f' * w_3$ =   _____

10. $\text{VPYy} = yo - f' * w_4$ =   _____

11. $\text{VPZx} = xo - f' * w_5$ =   _____

12. $\text{VPZy} = yo - f' * w_6$ =   _____

13.6 Two Vanishing Points and Principal Point Method

The third major vanishing point can be determined if the photogrammetric analyst already knows the other two major vanishing points and the principal point. There are three procedures listed in this section for determining the unknown major vanishing point. The image-coordinate system origin is some other point than the principal point. Determine the values in the order given, see Example E-12 in the appendix.

Procedure No. 1 - VPX, VPY, and pp are known, find VPZ

1. $dx_1 = VPXx - xo$
2. $dx_2 = VPYx - xo$ = ...
3. $dy_1 = VPXy - yo$ = ...
4. $dy_2 = VPYy - yo$ = ...
5. $w_1 = (VPXy * dy_2) + (VPXx * dx_2)$ = ...
6. $w_2 = (VPYy * dy_1) + (VPYx * dx_1)$ = ...
7. $w_3 = (w_2 * dy_1) - (w_1 * dy_2)$ = ...
8. $w_4 = (w_1 * dx_2) - (w_2 * dx_1)$ = ...
9. $w_5 = (dy_1 * dx_2) - (dy_2 * dx_1)$ = ...
10. $VPZx = w_3 / w_5$ =   _____
11. $VPZy = -w_4 / w_5$ =   _____

Procedure No. 2 - VPY, VPZ, and pp are known, find VPX

1. $dx_1 = -VPYx + xo$ = ...
2. $dx_2 = VPZx - VPYx$ = ...
3. $dy_1 = -VPYy + yo$ = ...
4. $dy_2 = VPZy - VPYy$ = ...
5. $w_1 = (yo * dy_2) + (xo * dx_2)$ = ...
6. $w_2 = (VPZy * dy_1) + (VPZx * dx_1)$ = ...
7. $w_3 = (w_1 * dy_1) - (w_2 * dy_2)$ = ...
8. $w_4 = (w_1 * dx_1) - (w_2 * dx_2)$ = ...
9. $w_5 = (dy_1 * dx_2) - (dy_2 * dx_1)$ = ...
10. $VPXx = w_3 / w_5$ =   _____
11. $VPXy = -w_4 / w_5$ =   _____

Procedure No. 3 - VPX, VPZ, and pp are known, find VPY

1. $dx_1 = VPZx - VPXx$ = ...
2. $dx_2 = xo - VPXx$ = ...
3. $dy_1 = VPZy - VPXy$ = ...
4. $dy_2 = yo - VPXy$ = ...
5. $w_1 = (VPZy * dy_2) + (VPZx * dx_2)$ = ...
6. $w_2 = (yo * dy_1) + (xo * dx_1)$ = ...
7. $w_3 = (w_1 * dy_1) - (w_2 * dy_2)$ = ...
8. $w_4 = (w_1 * dx_1) - (w_2 * dx_2)$ = ...
9. $w_5 = (dy_1 * dx_2) - (dy_2 * dx_1)$ = ...
10. $VPYx = w_3 / w_5$ = _____
11. $VPYy = -w_4 / w_5$ = _____

13.7 Conclusion

Five methods — two lines, segmented line, orthogonal line, interior parametric values, and two **VP**s and **pp** — were presented on how to determine the image coordinates of a vanishing point. Of these methods, the two-lines method will probably be used the most because it has such versatility. Also, depending upon the convergent angle of the two-lines, it can be the most inaccurate. Still, the two line method is the easiest to verify if there are more than two lines. And, for those with programmable calculators or desktop computers, there is a least squares simultaneous solution of three or more lines to a vanishing point. A description of this procedure is in the Appendix.

DIMENSIONAL ANALYSIS THROUGH PERSPECTIVE

Chapter Fourteen
Perspective Rotation Matrix

14.1 Introduction

There are three rotation angles involved in the photogrammetric analysis of single-image perspective. These angles have been defined in Chapter Three. The angles are: azimuth (**a**), tilt (**t**), and swing (**s**). These angles define the attitude of the principal ray in image- and object-space coordinate systems. There are many methods to be used in computing the angles and the corresponding rotation matrix [**R**]. In this Chapter three methods will be presented for determining [**R**]. The first method uses the azimuth, tilt, and swing angles in decimal degrees. The second method uses the **pp, f', a,** and one major vanishing point. The third method uses the procedures of the second method plus all three major vanishing points. The last item discussed in this chapter is how to determine if the matrix itself is in error.

14.2 Three Rotation Angles (a, t, and s) Method

For the purposes of calculation, the photogrammetric analyst should be aware of accuracy and significant figures. If the following trigonometric functions are selected from a text, each sine and cosine value should be carried out to at least six places to the right of the decimal. To carry the values out to nine places would be better, and if a pocket calculator is used, values should be extended to the maximum range of the calculator. The following values should be determined by whatever method is used.

Rotation Matrix from Azimuth, Tilt and Swing (E-17)

1.	$SA = \sin(a)$	=	....
2.	$CA = \cos(a)$	=	....
3.	$ST = \sin(t)$	=	....
4.	$CT = \cos(t)$	=	....
5.	$SS = \sin(s)$	=	....
6.	$CS = \cos(s)$	=	....
7.	$w_1 = -SA * CT$	=	....
8.	$w_2 = -CA * CT$	=	....
9.	$r_{11} = (w_1 * SS) - (CA * CS)$	=	_____

10. $r_{12} = (w_2 * SS) + (SA * CS)$ =   _____

11. $r_{13} = -ST * SS$ =   _____

12. $r_{21} = (w_1 * CS) + (CA * SS)$ =   _____

13. $r_{22} = (w_2 * CS) - (SA * SS)$ =   _____

14. $r_{23} = -ST * CS$ =   _____

15. $r_{31} = -ST * SA$ =   _____

16. $r_{32} = -ST * CA$ =   _____

17. $r_{33} = CT$ =   _____

14.3 Rotation Matrix from pp, f', a, and Major VP

This method consists of three procedures since there are three major vanishing points. The origin of the image-coordinate system is some point other than the principal point. The values should be calculated in the order given.

[R] from pp, f',a, and one VP (E-18)
Procedure No. 1 - VP = VPX

1. $w_1 = VPXx - xo$ = ...

2. $w_2 = VPXy - yo$ = ...

3. $VRx = (w_1^2 + w_2^2 + f'^2)^{1/2}$ = ...

4. $r_{11} = w_1 / VRx$ =   _____

5. $r_{21} = w_2 / VRx$ =   _____

6. $r_{31} = -f' / VRx$ =   _____

7. $r_{32} = r_{31} / \tan(a)$ =   _____

8. $r_{33} = (1.0 - r_{31}^2 - r_{32}^2)^{1/2}$ =   _____
   [if camera points upward, then $r_{33} = -r_{33}$]

9. $w_3 = (r_{21} * r_{33}) - (r_{11} * r_{31} * r_{32})$ = ...

10. $w_4 = r_{11}^2 + r_{21}^2$ = ...

11. $r_{22} = w_3 / w_4$ =   _____

12. $w_5 = (r_{11} * r_{22}) - r_{33}$ = ...

13. $r_{12} = w_5 / r_{21}$ =   _____

14. $w_6 = (r_{11} * r_{31}) + (r_{12} * r_{32})$ = ...

15. $w_7 = (r_{21} * r_{31}) + (r_{22} * r_{32})$ = ...

16. $r_{13} = -w_6 / r_{33}$ =   _____

17. $r_{23} = -w_7 / r_{33}$ =   _____

[R] from pp, f', a, and one VP (E-18)
Procedure No. 2 - VP = VPY

1. $w_1 = VPXx - xo$ = ...
2. $w_2 = VPXy - yo$ = ...
3. $VRy = (w_1^2 + w_2^2 + f'^2)^{1/2}$ = ...
4. $r_{12} = w_1 / VRy$ =   _____
5. $r_{22} = w_2 / VRy$ =   _____
6. $r_{32} = -f' / VRy$ =   _____
7. $r_{31} = r_{32} / \tan(a)$ =   _____
8. $r_{33} = (1.0 - r_{31}^2 - r_{32}^2)^{1/2}$ =   _____
      [if camera points upward, then r_{33}-r_{33}]
9. $w_3 = (r_{31} * r_{12}) - (r_{22} * r_{32} * r_{33})$ = ...
10. $w_4 = r_{12}^2 + r_{22}^2$ = ...
11. $r_{23} = w_3 / w_4$ =   _____
12. $w_5 = (r_{12} * r_{29}) - r_{31}$ = ...
13. $r_{13} = w_5 / r_{22}$ =   _____
14. $w_6 = (r_{12} * r_{32}) + (r_{13} * r_{33})$ = ...
15. $w_7 = (r_{22} * r_{32}) + (r_{23} * r_{33})$ = ...
16. $r_{13} = -w_6 / r_{31}$ =   _____
17. $r_{21} = -w_7 / r_{31}$ =   _____

[R] from pp, f', a, and one VP (E-18)
Procedure No. 3 - VP = VPZ

1. $w_1 = VPZx - xo$ = ...
2. $w_2 = VPZy - yo$ = ...
3. $VRz = (w_1^2 + w_2^2 + f'^2)^{1/2}$ = ...
4. $r_{13} = -w_1 / VRz$ =   _____
5. $r_{23} = -w_2 / VRz$ =   _____
6. $r_{33} = f' / VRz$ =   _____
      [if camera points up, then $r_{13} = -r_{13}$, $r_{23} = -r_{23}$, $r_{33} = -r_{33}$]
7. $w_3 = r_{13}^2 + r_{23}^2$ = ...
8. $w_4 = -(w_3)^{1/2}$ = ...
9. $r_{31} = w_4 * \sin(a)$ =   _____
10. $r_{32} = w_4 * \cos(a)$ =   _____

11. $w_5 = (r_{32} * r_{13}) - (r_{23} * r_{31} * r_{33})$ $=$ ....

12. $r_{21} = w_5 / w_3$ $=$ _____

13. $w_6 = (r_{21} * r_{13}) - r_{33}$ $=$ ....

14. $r_{11} = w_6 / r_{23}$ $=$ _____

15. $w_7 = (r_{11} * r_{31}) + (r_{13} * r_{33})$ $=$ ....

16. $w_8 = (r_{21} * r_{31}) + (r_{23} * r_{33})$ $=$ ....

17. $r_{12} = -w_6 / r_{32}$ $=$ _____

18. $r_{22} = -w_7 / r_{32}$ $=$ _____

14.4 The pp, f', and Three Major VPs Method

The origin of the image-coordinate system is a point other than the principal point. The values should be calculated in the order given.

pp, f', VPX, VPY, and VPZ Method (E-19)

1. $w_1 = VPXx - xo$ $=$ ....

2. $w_2 = VPYx - xo$ $=$ ....

3. $w_3 = VPZx - xo$ $=$ ....

4. $w_4 = VPXy - yo$ $=$ ....

5. $w_5 = VPYy - yo$ $=$ ....

6. $w_6 = VPZy - yo$ $=$ ....

7. $VRx = (w_1^2 + w_4^2 + f'^2)^{1/2}$ $=$ ....

8. $VRy = (w_2^2 + w_5^2 + f'^2)^{1/2}$ $=$ ....

9. $VRz = (w_3^2 + w_6^2 + f'^2)^{1/2}$ $=$ ....

10. $r_{11} = w_1 / VRx$ $=$ _____

11. $r_{21} = w_4 / VRx$ $=$ _____

12. $r_{31} = -f' / VRx$ $=$ _____

13. $r_{12} = w_2 / VRy$ $=$ _____

14. $r_{22} = w_5 / VRy$ $=$ _____

15. $r_{32} = -f' / VRy$ $=$ _____

16. $r_{13} = w_3 / VRz$ $=$ _____

17. $r_{23} = w_6 / VRz$ $=$ _____

18. $r_{33} = -f' / VRz$ $=$ _____

Note: If the value of w_6 is negative then: $r_{13} = -r_{13}$, $r_{23} = -r_{23}$, and $r_{33} = -r_{33}$

14.5 Checking the Rotation Matrix for Errors

Everyone interested in photogrammetric analysis should obtain a college outline series on matrices and vectors and review orthogonality. The photogrammetric rotation matrix is an orthogonal matrix, and has specific properties that will allow error to be checked. If the matrix is mathematically correct, it will be possible to compute the following:

1. $r_{11}^2 + r_{12}^2 + r_{13}^2 = 1.000000$ = ...
2. $r_{21}^2 + r_{22}^2 + r_{23}^2 = 1.000000$ = ...
3. $r_{31}^2 + r_{32}^2 + r_{33}^2 = 1.000000$ = ...
4. $r_{11}^2 + r_{21}^2 + r_{31}^2 = 1.000000$ = ...
5. $r_{12}^2 + r_{22}^2 + r_{32}^2 = 1.000000$ = ...
6. $r_{13}^2 + r_{23}^2 + r_{33}^2 = 1.000000$ = ...
7. $r_{11}{}^*r_{21} + r_{12}{}^*r_{22} + r_{13}{}^*r_{23} = 0.000000$ = ...
8. $r_{11}{}^*r_{31} + r_{12}{}^*r_{32} + r_{13}{}^*r_{33} = 0.000000$ = ...
9. $r_{21}{}^*r_{31} + r_{22}{}^*r_{32} + r_{23}{}^*r_{33} = 0.000000$ = ...
10. $r_{11}{}^*r_{12} + r_{21}{}^*r_{22} + r_{31}{}^*r_{32} = 0.000000$ = ...
11. $r_{11}{}^*r_{13} + r_{21}{}^*r_{23} + r_{31}{}^*r_{33} = 0.000000$ = ...
12. $r_{12}{}^*r_{13} + r_{22}{}^*r_{23} + r_{32}{}^*r_{33} = 0.000000$ = ...

From the 12 values calculated above, it is possible to determine which element is in error. For example, if equation 1. does not equal 1.0 to the accuracy required, and equation 4. is also incorrect, and equations 2, 3, 5, and 6. are correct, then the error must be in r11 because it is only common to equations 1 and 4. One of the first matrix elements to check is r_{33}. Look through the equations to determine if r_{33} is in error. The purpose for this is that r_{33} is the cosine value of the tilt angle (see Section 14.2), and the one matrix value directly attributed to a rotation angle. If r_{33} is correct, then check r_{31} and r_{32}. These two matrix elements involve the azimuth and tilt angles, and if it has been proven that the tilt is correct, the error in these elements must be in the azimuth. Likewise, check the r_{13} and r_{23} matrix elements. These elements involve the tilt and swing angles. These checks are tedious if calculated without the support of programmable calculators (or desktop computers), but they certainly support the reliability of the work. Again we point out that it is important to double check the analysis, and determine which values can be adjusted to provide verifiable results.

14.6 Conclusion

The camera rotation matrix [**R**] can be computed from several alternative starting points. This chapter has shown how [**R**] can be determined from the three camera angles of azimuth, tilt, and swing. We have also shown how to compute [**R**] from azimuth, effective focal length, and a major vanishing point. (The second method is slightly more complicated than the first, as might be expected.) Finally, we have shown how to compute [**R**] from the principal point, effective focal length, and the three major vanishing points. Other permutations of given information can be imagined from which [**R**] can be determined. We have discussed three common starting points with the idea that a photogrammetric analyst will be able to extend these methods to other situations.

DIMENSIONAL ANALYSIS THROUGH PERSPECTIVE

Chapter Fifteen
Focal Length

15.1 Introduction

The focal length (principal distance) can be any one of three values: the infinity calibrated focal length, the lens nominal focal length, or the effective focal length at the time of collection. For close-range photogrammetry, theoreticians stipulate that a calibrated focal length is required, while the nominal focal length is usually what is provided. What is required is the effective focal length (f') at the time of the collection (exposure). Without the f', the photogrammetric analyst is starting his analysis with a built-in error that may not be acceptable. There are times when it is possible to calculate an effective focal length from the image geometry. For a non-calibrated, off-the-shelf cameras, whose lens was just taken off another camera, and may have been positioned correctly on the camera body, the effective focal length calculated from the geometry is the best approximation of the effective focal length possible. In two-point perspective, the effective focal length is the perpendicular distance from the xy true horizon line (**THLxy** or **THLXY**) through the principal point to the image graphical position of the camera station. In three-point perspective the effective focal length is identically found from each of the true horizon lines (see Figure 15-1). In some perspective solutions, there is a need to have a known focal length, and the nominal focal length will allow the analyst to determine a solution. A better solution will be determined if the photogrammetric analyst determines the effective focal length from the geometry of the image.

15.2 Two-Point Perspective Focal Length

For all practical purposes, any perspective imagery may be considered two-point perspective when the tilt angle is in the range of 87° to 93° (that plus or minus three degrees from the horizontal). One of the major reasons for selecting this range is that object-space vertical lines at this tilt angle will appear vertical in image-space. Also, at angles of three degrees or less there is a potential for more human error in measuring points on the vertical image line (small angular displacement cause large position errors). Another way of stating $\longrightarrow$ continued on next page

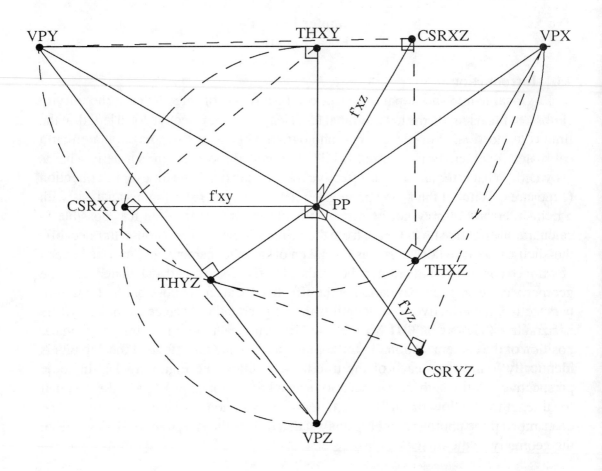

Figure 15-1. Three-Point Perspective Focal Length

this is that any small error in selecting and measuring an image point on the image vertical lines would be greatly magnified in the area of the intersection point (vanishing point). It is better to have your errors where you can determine error magnitude and make use of them, than it is to have them where you cannot determine what they are and cannot use them. (And somewhere out there in that vast jungle of mathematics is a mathematician ~~that~~ *who* will be able to determine, for angles less than three degrees, that the tangent function is almost the same as the sine function, so you might as well use vertical lines, and so on.) Someone once said, "If it works and it's not broken, then don't fix it." The analytical analysis of two-point perspective, in that angle range, works!

The best two-point solution for the effective focal length is to determine the coordinates of two of the major vanishing points, locate the **pp** on the **THLxy** as the intersection of the full frame format diagonals, and then determine the **f'** as the distance from the **pp** to the semicircle with a diameter of the distance between the two major vanishing points (see Figure 4-3). There is no need to provide a lengthy analytical solution for this if the **pp** is the origin of the image coordinate system, and the x-axis is coincident with the **THLxy**, because the **f'** is the absolute value of the y coordinate for the image graphical camera station. Of course, if you cannot find the principal point because the imagery does not have full format, you may have a problem. The next best solution is to use one of the diagonal methods discussed in Chapter Six. The diagonal method will not only provide a solution to the **f'**, but will also provide all of the required Phase One parametric values.

15.3 Three-Point Perspective Focal Length

When we look at the graphic illustration of the three major vanishing points, we are looking at a perspective triangle. This perspective triangle is the base for the perspective pyramid (or tetrahedron). The image plane is in the base of the pyramid, and the camera station is at the apex of the pyramid. The perspective pyramid provides much more geometry to describe the focal length than this description. The perspective triangle (base of the pyramid) is composed of the three horizon traces (XY, XZ, YZ), and these horizon traces are each the hypotenuse of a right triangle where the point at the right angle is the camera station (the apex of the pyramid). Also, there is a line that is perpendicular to each horizon trace that intersects with the vanishing point opposite the horizon trace. Let's call these last lines horizon trace principal lines (**HTPL**). Each **HTPL** is the hypotenuse of a right triangle, a principal line right triangle (**PLRT**). Care to guess where the right angle point of each of these PLRTs meet? Correct, at the apex of the perspective pyramid — the camera station. Each **PLRT** has a common line with the other **PLRT**s, the **f'**. The **PLRT** that contains the **THxy**, **pp**, and **VPZ** is also called the principal plane. To save us from further confusion this geometry is illustrated in Figures 15-2 through 15-4. In Figure 15-2 the complete perspective pyramid is shown. In Figure 15-3 only the **PLRT**s are shown, and in Figure 15-4, only the **PLRT** of the principal plane is shown. The **PLRT** in Figure 15-4 is also shown as a rotated plane about the principal line (**TH-VPZ**) in the perspective triangle. It is from this geometry that **f'** will be determined. Since there are three **PLRT**s, there are three solutions to **f'**, and all three solutions will be provided. Some will insist that all three solutions be solved to ensure accuracy. We do not insist on this, however it is always a good procedure to check your work and verify results in as many different ways as your time and need allows. We can almost guarantee that if you don't, an error will have occurred, and if you do there will not be any significant difference in the values (Murphy is always with us).

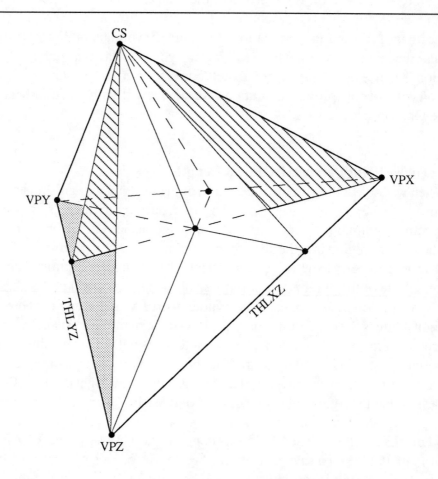

Figure 15-2. Three-point Perspective Pyramid

The geometry of the perspective pyramid is fixed once you have determined three major vanishing points, and unless you make a blunder in entering data or working the calculations, the only difference in the following solutions for **f'** will be because of your calculator. In the following solutions it is assumed that the principal point is not the origin of the image-coordinate system.

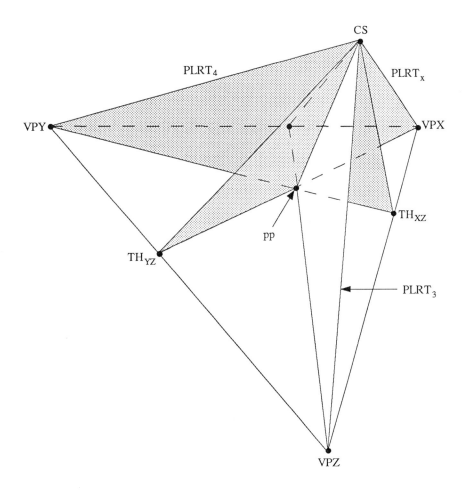

Figure 15-3. Principal Line Right Triangles

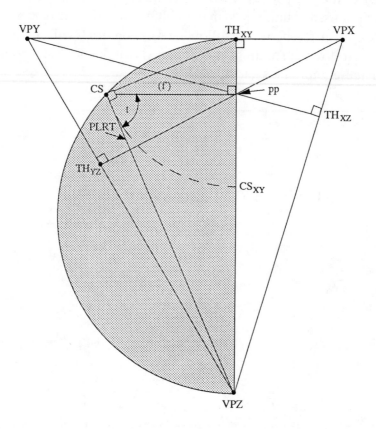

Figure 15-4. Principal Line Right Triangle through VPZ

f' using pp, VPX, and VPY (E-20)

1. $w_1 = VPXx - xo$ $=$ ....
2. $w_2 = VPYx - xo$ $=$ ....
3. $w_3 = VPXy - yo$ $=$ ....
4. $w_4 = VPYy - yo$ $=$ ....
5. $w_5 = (w_1 * w_2) + (w_3 * w_4)$ $=$ ....
6. $\mathbf{f'} = (|w_5|)^{1/2}$ $=$ _____

f' using pp, VPX, and VPZ (E-20)

1. $w_1 = VPYx - xo$ = ...
2. $w_2 = VPZx - xo$ = ...
3. $w_3 = VPYy - yo$ = ...
4. $w_4 = VPZy - yo$ = ...
5. $w_5 = (w_1{}^*w_2) + (w_3{}^*w_4)$ = ...
6. $\mathbf{f'} = (|w_5|)^{1/2}$ =   _____

f' using pp, VPY, and VPZ (E-20)

1. $w_1 = VPYx - xo$ = ...
2. $w_2 = VPZx - xo$ = ...
3. $w_3 = VPYy - yo$ = ...
4. $w_4 = VPZy - yo$ = ...
5. $w_5 = (w_1{}^*w_2) + (w_3{}^*w_4)$ = ...
6. $\mathbf{f'} = (|w_5|)^{1/2}$ =   _____

15.4 Conclusion

This chapter has shown how to compute the effective focal length analytically from the principal point and two major vanishing points, for both two-point and three-point perspective images. In three-point perspective, the perspective pyramid was used to aid visualization of the result.

DIMENSIONAL ANALYSIS THROUGH PERSPECTIVE

Chapter Sixteen
Azimuth, Tilt, and Swing

16.1 Compute the Azimuth, Tilt, and Swing Angles

Rotation angles were discussed in Chapter Three, and now present the methods for determining those angles. When photography was first used to collect images of large areas, the best camera station available was in a balloon, up above the area of interest. Science observed that by pointing the camera straight down, the best geometry was obtained. (Science, that's Murphy's playroom, but for a very good description of the development of photogrammetry read the Manual of Photogrammetry, Fourth Edition, ASPRS). Soon the science of cartography took over and photogrammetry was just a supporting science. Consequently, the azimuth angle is about the object space Z-axis, the tilt angle is a rotation about a line parallel to the horizontal horizon line through the camera station (lens), and the swing angle is a rotation of the imagery about the principal ray (principal line, focal length, whatever).

When photogrammetrists are as good as they think they are these angles can be approximated with useful accuracy. We already know that if the imagery is two point perspective the tilt angle is very close to 90° (not zero degrees — remember they took the first photographic exposures for cartographic purposes looking straight down, because the object space was more interesting). It is also easy to determine that if the XY horizon line parallels the horizontal edge of the image format the swing angle is 180°. The azimuth angle is the only angle that is not easy to approximate, but there are occasions when the geometry for azimuth allows the photogrammetric analyst to make acceptable approximations.

A major point about approximations is to use every possible means to double check the results. To check the work is fairly simple. After many grueling hours of collecting and analyzing the data determine if the locus of calculated object-space points for certain geometric forms are true. Are rectangular corners square, is a circle round, and are perpendicular planes perpendicular? Is the scale constant along each axis? Do all the known dimensions (lines and angles) agree with calculated values within accuracy requirements? Do the nine elements of the rotation matrix meet prescribed mathematical checks and balances given in Chapter Fifteen (15.5). The photogrammetric analysis could be wrong because of an error in a rotation angle or angles, and it might not be initially obvious, but these errors will eventually become known. Dr. Murphy will not let errors stay unnoticed for long.

The methods for obtaining the Phase One rotation angles presented in this manual are straightforward and simple. First, given the rotation matrix it is possible to calculate the rotation angles, and second, given the other Phase One parametric values it is possible to calculate the rotation angles. These procedures are described in the following paragraphs.

16.2 Azimuth, Tilt, and Swing Given [R]

Remember the rotation matrix contains nine elements (Section Seven). These are the direction cosine values for a particular vector, the principal ray. Therefore, it is possible that the rotation matrix was developed from some other combination of useful angles (omega, phi, and kappa; pitch, roll, and yaw; and so on). In any case it is the same rotation matrix of nine numbers. Your first error check is that if the matrix has a single element that has a larger absolute value than one, the matrix is in error. Given a correct rotation matrix the rotation angles are determined as follows:

Azimuth, Tilt, and Swing from [R] (E-21)

1. $w_1 = r_{31} / r_{32}$ = ...
2. $w_2 = r_{13} / r_{23}$ = ...
3. $a = Tan^{-1}(|w_1|)$ [0° to 90°]   =   _____
4. $t = Cos^{-1}(r_{33})$ [if r_{33} neg.]   =   _____
5. $t = Sin^{-1}(w_3)$ [if r_{33} pos.]   =   _____
        [where t = 0° to 180°]
6. $s = Tan^{-1}(w_2) + 180°$ [90° to 270°]   =   _____

It should also be understood that Step 5 gives a more reliable answer for tilt when the camera is close to being horizontal (90°) and Step 6 provides a more reliable answer for tilt when the camera is pointing at the nadir or the zenith (0° or 180°). So even though the negative value of r_{33} indicates an angle between 0° and 90°, discretion should be used by the photogrammetrist when calculating tilt. Balloon imagery would tend to use

Step 6, where as buttonhole imagery would tend to use equation 5 to solve for tilt. The swing angle has some cautions also, in that for angles from 90° to 180° the w_2 value is negative and for angles from 180° to 270° the w_2 values is positive. This means look at your imagery and make a decision about which equation to use.

16.3 Azimuth, Tilt, and Swing Given Phase One Values

We are not too sure why anyone would want to calculate the rotation angles before you calculated the rotation matrix, except it is easier to record three numbers and calculate the rotation matrix whenever it is needed. The saying is, "necessity is the mother of invention" or should it be, "laziness is the mother of invention"? In any case the procedure for using this method is as follows:

Using pp, f', VPX, VPY, and VPZ (E-22)

1. $w_1 = VPXx - xo$ = ...
2. $w_2 = VPYx - xo$ = ...
3. $w_3 = VPZx - xo$ = ...
4. $w_4 = VPXy - yo$ = ...
5. $w_5 = VPYy - yo$ = ...
6. $w_6 = VPZy - yo$ = ...
7. $VRx = (w_1^2 + w_4^2 + f'^2)^{1/2}$ = ...
8. $VRy = (w_2^2 + w_5^2 + f'^2)^{1/2}$ = ...
9. $VRz = (w_3^2 + w_6^2 + f'^2)^{1/2}$ = ...
10. $w_7 = VRy / VRx$ = ...
11. $w_8 = f' / VRz$ = ...
12. $w_9 = (w_3^2 + w_6^2)^{1/2} / VRz$ = ...
13. $w_{10} = w_3 / w_6$ = ...
14. $a = Tan^{-1}(|w_7|)$ =   _____
15. $t = Cos^{-1}(w_8)$ [if w_6 is neg.]   =   _____
16. $t = Sin^{-1}(w_9)$ [if w_6 is pos.]   =   _____
17. $s = Tan^{-1}(w_{10}) + 180°$ =   _____

The same cautions apply here as listed in paragraph 16.1. For near horizontal imagery equation 15 will provide the more reliable values of the tilt angle. The value of w_6

indicates camera up or down by the location of VPZ. If the origin of the image coordinate system is the principal point then the sign of VPZ is the indicator.

16.4 Conclusion

The computation of azimuth, tilt, and swing can be done readily from the rotation matrix [**R**] or given other Phase One parameters values (VPs, **pp**, and **f'**). Generally the rotation matrix is obtained prior to camera rotation angles, but this order of information extraction is not absolutely necessary.

DIMENSIONAL ANALYSIS THROUGH PERSPECTIVE

Chapter Seventeen
Phase Two Parametric Procedures

17.1 Introduction

Procedures of calculating Phase Two values apply to many single-image perspective situations. We would not be so bold as to say they apply to all situations, first because we have not experienced them all and secondly because, if we thought we had, Dr. Murphy would deliberately bring another perspective procedure to our attention. The perspective geometries encountered in single-image perspective are varied. In this manual an attempt has been made to present the construction for understanding the geometry. We make no claims to having developed all of these geometric procedures; some we derived, some were demonstrated to us, and some have been around so long it seems they have always been in the public domain. In most cases we feel the procedures are unique enough to warrant using them, and the rest could not be left out in good conscience.

The photogrammetric analyst should become familiar with the procedures of each situation and select from them the best methods for his own bag (shoe box) of perspective analysis solutions. It should be obvious that not all of the solutions will apply to your particular tasks, however a mixture of available procedures allows one to develop a capability with single-photo perspective. Also, tasks without a challenge are not fun, and you should enjoy what you do.

17.2 Projective Equations

A discussion of the projective equations is given in paragraph 12.1 of Chapter Twelve. These are the most important equations, and the photogrammetric analyst should understand how they were developed and how they are used.

17.3 Phase Two Methods

Presented in the following chapters are five analytical methods and associated procedures, which are used in the second phase of the photogrammetric solutions to perspective imagery. These methods are:

1. Two point perspective - plan view
2. Two point perspective - elevation views
3. Three point perspective - camera station
4. Three-point perspective - object-space dimensions
5. Three-point perspective - object-space coordinates.

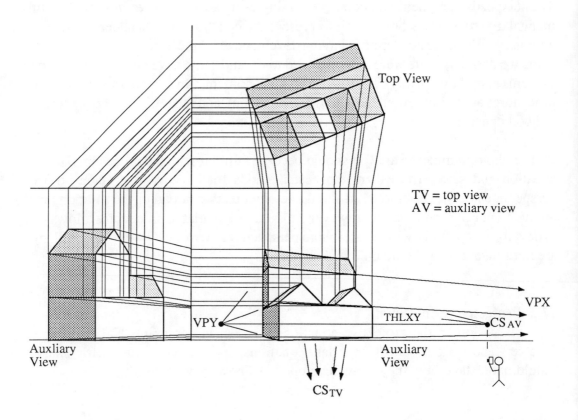

Figure 17-1. Two-Point Perspective from Orthogonal Lines

Figures 17-1 and 17-2 are graphical examples of two-point perspective. Figure 17-1 illustrates how two-point perspective is obtained from two orthographic views. The purpose of this figure is just to remind the photogrammetric analyst that two-point perspective is unique and has unique properties. On the other hand, Figure 17-2 displays the graphical procedures for horizontal and vertical dimensions. A reminder to the photogrammetric analyst is that when working a graphical solution it is important that you keep very accurate and timely notes. The same procedure applies to analytical procedures - keep accurate and timely notes.

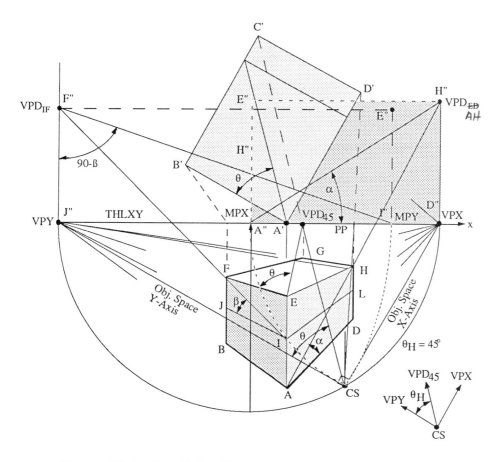

Figure 17-2. Two-Point Perspective with Plan and Elevations

The required parametric inputs are the Phase One values, plus specific object-space information. As stated in section 17.2, some of the math is designed around the projective equations, and some around the two dimensional geometry of the graphical solution. The reason is to provide the user with checks on a graphical solution (if so desired) and this manual is not meant to be a mathematical treatise on photogrammetry. Proof of solutions and development of each math model are not provided deliberately. It is up to the readers to prove a method, if they are so inclined.

The data required for these methods should be placed on a form that can be generated to suit the photogrammetric analyst's needs. A reminder concerning the form: be sure to include all units for given, measured, and calculated values. They may be needed later as reference data, or to prove how the solution was obtained.

17.4 Conclusion

The Phase Two parametric values are expressed as parts of the projective equations, and standard geometry form.The procedures for using, or finding, the Phase Two valves are described in five methods. The procedures can be used graphically, analytically, or as a combination of graphical and analytical procedures. The Phase One parameteric valves are required, and the analyst should keep detailed notes of the procedure and the measured points used in obtaining the Phase Two values.

DIMENSIONAL ANALYSIS THROUGH PERSPECTIVE

Chapter Eighteen
Camera-Station Computations

18.1 Introduction

In the analytical solution of a three-point perspective image the camera station is the pass point for scale. A known dimension in object space is selected so that it is possible to assign coordinates to both ends of the dimension. The best method for doing this is to select a dimension that is parallel to one of the major axes and assign one end to be the origin of the object-space coordinate system or any other arbitrary set of coordinates. It may be that it is desirable to have the camera-station coordinates be positive numbers; therefore, assign one end of the known dimension the coordinates 1000.0, 1000.0, 1000.0, in whatever units are usable. Of course, that assumes that the range from the camera station to the dimension is inside the boundaries set up by those coordinates.

The methods given in this chapter are those to calculate a single camera-station coordinate, two camera-station coordinates, or all three camera station coordinates. The parametric values required are the Phase One parameters and the image coordinates for all measured points. To establish scale, and the object-space coordinate system, there is a need for a dimension parallel to one of the major axis planes. An established coordinate system can be used if the coordinates of the ends of the dimension are known, or you can use the dimension to establish the object-space coordinate system. In either case the coordinates of the end points of the dimension are required to calculate one camera-station coordinate. With this information, and the one camera station coordinate, it is possible to calculate the remaining two camera-station coordinates.

18.2 One Camera-Station Coordinate

The solutions all depend upon measured image values with the image coordinate origin not being the principal point. Compute the following values in the order given, see Example Twenty-three (E-23) in the Appendix.

Method for Calculating a CS Coordinate

The procedures for determining Xc, Yc, or Zc are identical until Step 11, at which time the plane of the given distance determines the completion of the solution.

Given a distance parallel to a major plane: DXY, DXZ, DYZ

The end point of the known dimension with the known object-space coordinate is point number 1 in the solution. Using whatever means available, the image points have been measured and recorded with known units. All of the Phase One parametric values for the image are known.

1.	w_1	=	$x_1 - xo$	=	....
2.	w_2	=	$x_2 - xo$	=	....
3.	w_3	=	$y_1 - yo$	=	....
4.	w_4	=	$y_2 - yo$	=	....
5.	w_5	=	$r_{11}*w_1 + r_{21}*w_3 - r_{31}*f'$	=	....
6.	w_6	=	$r_{12}*w_1 + r_{22}*w_3 - r_{32}*f'$	=	....
7.	w_7	=	$r_{13}*w_1 + r_{23}*w_3 - r_{33}*f'$	=	....
8.	w_8	=	$r_{11}*w_2 + r_{21}*w_4 - r_{31}*f'$	=	....
9.	w_9	=	$r_{12}*w_2 + r_{22}*w_4 - r_{32}*f'$	=	....
10.	w_{10}	=	$r_{13}*w_2 + r_{23}*w_4 - r_{33}*f'$	=	....

Case One: the given distance is parallel to the XY plane:

11.	w_{11}	=	w_5 / w_7	=	....
12.	w_{12}	=	w_6 / w_7	=	....
13.	w_{13}	=	w_8 / w_{10}	=	....
14.	w_{14}	=	w_9 / w_{10}	=	....
15.	w_{15}	=	$w_{13} - w_{11}$	=	....
16.	w_{16}	=	$w_{14} - w_{12}$	=	....
17.	w_{17}	=	$(w_{15}^2 + w_{16}^2)^{1/2}$	=	....

The general rule is if $y_1 <$ VPXy and $y_1 <$ VPYy then $w_{17} = - w_{17}$. Stated by definition Z_1 is greater than Zc $(Z_1 > Zc)$, if not then $w_{17} = - w_{17}$. $Z_1 < Zc$ is when the imaged origin point (point 1) is below the horizon line.

18.   Given X_1: Xc   = $X_1 - (w_{11} * (DXY/w_{17}))$ =   _____
        Given Y_1: Yc   = $Y_1 - (w_{12} * (DXY/w_{17}))$ =   _____
        Given Z_1: Zc   = $Z_1 - (DXY / w_{17})$ =   _____

Case Two: the given distance is parallel to the XZ plane:

11. w_{11} = w_5 / w_6 = ...
12. w_{12} = w_7 / w_6 = ...
13. w_{13} = w_8 / w_9 = ...
14. w_{14} = w_{10} / w_9 = ...
15. w_{15} = $w_{13} - w_{11}$ = ...
16. w_{16} = $w_{14} - w_{12}$ = ...
17. w_{17} = $(w_{15}^2 + w_{16}^2)^{1/2}$ = ...
By definition, $Y_1 > Yc$: if not then $w_{17} = -w_{17}$

18.   Given X_1: Xc   = $X_1 - (w_{11} * (DXZ/w_{17}))$ =   _____
        Given Y_1: Yc   = $Y_1 - (DXZ / w_{17})$ =   _____
        Given Z_1: Zc   = $Z_1 - (w_{12} * (DXZ/w_{17}))$ =   _____

Case Three: the given distance is parallel to the YZ plane:

11. $w_{11} = w_6 / w_5$ = ...
12. $w_{12} = w_7 / w_5$ = ...
13. $w_{13} = w_9 / w_8$ = ...
14. $w_{14} = w_{10} / w_8$ = ...
15. $w_{15} = w_{13} - w_{11}$ = ...
16. $w_{16} = w_{14} - w_{12}$ —   ....
17. $w_{17} = (w_{15}^2 + w_{16}^2)^{1/2}$ = ...
By definition, $Y_1 > Yc$: if not then $w_{17} = -w_{17}$

18.   Given X_1: Xc   = $X_1 - (w_{11} * (DXZ/w_{17}))$ =   _____
        Given Y_1: Yc   = $Y_1 - (DXZ / w_{17})$ =   _____
        Given Z_1: Zc   = $Z_1 - (w_{12} * (DXZ/w_{17}))$ =   _____

18.3 Two Camera-Station Coordinates

The solutions all depend upon measured image values with the image coordinate origin not being the principal point. Compute the following values in the order given.

Method for Calculating Two CS Coordinates (E-24)

The procedures are identical until Step 6, where the given **CS** coordinate determines the solution. The end point of the known dimension, with the known object-space coordinates, is point number 1. All Phase One parametric values for the image are known.

1. w_1 = $x_1 - xo$ = ...
2. w_2 = $y_1 - yo$ = ...
3. w_3 = $r_{11}*w_1 + r_{21}*w_2 - r_{31}*f'$ = ...
4. w_4 = $r_{12}*w_1 + r_{22}*w_2 - r_{32}*f'$ = ...
5. w_5 = $r_{13}*w_1 + r_{23}*w_2 - r_{33}*f'$ = ...

Case One: where the known CS coordinate is Xc.

6. Yc = $Y_1 + (w_4 / w_3) * (Xc - X_1)$ =   _____
7. Zc = $Z_1 + (w_5 / w_3) * (Xc - X_1)$ =   _____

Case Two: where the known CS coordinate is Yc.

8. Xc = $X_1 + (w_3 / w_4) * (Yc - Y_1)$ =   _____
9. Zc = $Z_1 + (w_5 / w_4) * (Yc - Y_1)$ =   _____

Case Three: where the known CS coordinate is Zc.

10. Xc = $X_1 + (w_3 / w_5) * (Zc - Z_1)$ =   _____
11. Yc = $Y_1 + (w_4 / w_5) * (Zc - Z_1)$ =   _____

18.4 Three Camera Station Coordinates

These solutions depend upon measured image values with the image coordinate origin not being the principal point. Compute the following values in the order given.

Method for Calculating Three CS Coordinates (E-25)

The end points of the known dimension are numbered 1 and 2 (with point number 1 being the origin, if that solution is desired). The object-space coordinates of the points are known. All Phase One parametric values for the image are known.

1.	w_1	$=$	$x_1 - xo$	$=$	....
2.	w_2	$=$	$x_2 - xo$	$=$	....
3.	w_3	$=$	$y_1 - yo$	$=$	....
4.	w_4	$=$	$y_2 - yo$	$=$	....
5.	w_5	$=$	$r_{11}*w_1 + r_{21}*w_3 - r_{31}*f'$	$=$	....
6.	w_6	$=$	$r_{12}*w_1 + r_{22}*w_3 - r_{32}*f'$	$=$	....
7.	w_7	$=$	$r_{13}*w_1 + r_{23}*w_3 - r_{33}*f'$	$=$	....
8.	w_8	$=$	$r_{11}*w_2 + r_{21}*w_4 - r_{31}*f'$	$=$	....
9.	w_9	$=$	$r_{12}*w_2 + r_{22}*w_4 - r_{32}*f'$	$=$	....
10.	w_{10}	$=$	$r_{13}*w_2 + r_{23}*w_4 - r_{33}*f'$	$=$	....
11.	w_{11}	$=$	$-w_5 / w_9$	$=$	....
12.	w_{12}	$=$	w_8 / w_9	$=$	....
13.	w_{13}	$=$	w_6 / w_9	$=$	....
14.	w_{14}	$=$	$X_1 - X_2$	$=$	....
15.	w_{15}	$=$	$(w_{11}*Y_1)+(w_{12}*Y_2)+(w_{13}*w_{14})$	$=$	....
16.	w_{16}	$=$	$w_{11} + w_{12}$	$=$	....
17.	Yc	$=$	w_{15} / w_{16}	$=$	_____
18.	w_{17}	$=$	$(Yc-Y_1) * (w_5/w_6)$	$=$	....
19.	Xc	$=$	$X_1 + w_{17}$	$=$	_____
20.	w_{18}	$=$	$(Yc - Y_1) * (w_7/w_5)$	$=$	....
21.	Zc	$=$	$Z_1 + w_{18}$	$=$	_____

18.5 Conclusion

This chapter has enumerated methods for computing one or more camera-station coordinates from the Phase One parameters and a known object-space dimension parallel to an object-space coordinate plane. The methods each require the same information as input, but provide outputs that depend on the requirements of convenience and strength of geometry.

DIMENSIONAL ANALYSIS THROUGH PERSPECTIVE

Chapter Nineteen
Object-Space Dimensions

19.1 Introduction

The photogrammetric analyst wants to calculate a dimension on the object. The camera-station coordinates have been derived from procedures shown in Chapter Eighteen, and the end point of the line being sought is in a plane that is known or can be determined. Thus, with the appropriate camera-station and object-space coordinates, the length of the line can be determined. The procedures that can be used are listed in this section. The initial parametric values for this solution need to be calculated, and then the parametric values pertaining to the dimension. This is followed by the calculations for the dimension itself. The most obvious constraint is that each and every dimension calculated with this method must be parallel to one of the major planes.

The solution depends upon having the Phase One parametric values as known values, and the image measurements of the required points. The origin of the image coordinate system does not have to be the principal point. The computations are the same until equation 11, no matter which **CS** coordinate is known. Complete the calculations for the initial parametric values in the order given.

Initial Parametric Values (E-26)

1. $w_1 = x_1 - xo$ = ...

2. $w_2 = x_2 - xo$ = ...

3. $w_3 = y_1 - yo$ = ...

4. $w_4 = y_2 - yo$ = ...

5. $w_5 = r_{11}{}^{*}w_1 + r_{21}{}^{*}w_3 - r_{31}{}^{*}f'$ = ...

6. $w_6 = r_{12}{}^{*}w_1 + r_{22}{}^{*}w_3 - r_{32}{}^{*}f'$ = ...

7. $w_7 = r_{13}{}^{*}w_1 + r_{23}{}^{*}w_3 - r_{33}{}^{*}f'$ = ...

8. $w_8 = r_{11}{}^{*}w_2 + r_{21}{}^{*}w_4 - r_{31}{}^{*}f'$ = ...

9. $w_9 = r_{12}{}^{*}w_2 + r_{22}{}^{*}w_4 - r_{32}{}^{*}f'$ = ...

10. $w_{10} = r_{13}{}^{*}w_2 + r_{23}{}^{*}w_4 - r_{33}{}^{*}f'$ = ...

19.2 Given Xc Calculate the Dimension

Once the initial parametric values have been calculated, the following calculations are required for the dimension according to the major plane paralleled.

19.2.1   Given a CS Coordinate Calculate a Distance Parallel to the XY Plane

Compute the values in the order given.

11. $w_{11} = w_5 / w_7$ = ...

12. $w_{12} = w_6 / w_7$ = ...

13. $w_{13} = w_8 / w_{10}$ = ...

14. $w_{14} = w_9 / w_{10}$ = ...

15. $w_{15} = ((w_{13}-w_{11})^2+(w_{14}-w_{12})^2)^{1/2}$ = ...

Calculate the Distance Using Xc

16a. Dist. $= |(X_1 - Xc) * w_{15} / w_{11}|$ =   _____

Calculate the Distance Using Yc

16b. Dist. $= |(Y_1 - Yc) * w_{15} / w_{12}|$ =   _____

Calculate the Distance Using Zc

16c. Dist. $= |(Z_1 - Zc) * w_{15}|$ =   _____

To check the results compute the distance using at least two of the equations above (16a through 16c).

19.2.2   Given a CS Coordinate Calculate a Distance Parallel to the XZ Plane.
Compute the values in the order given.

11. $w_{11} = w_5 / w_6$ = ...

12. $w_{12} = w_7 / w_6$ = ...

13. $w_{13} = w_8 / w_9$ = ...

14. $w_{14} = w_{10} / w_9$ = ...

15. $w_{15} = ((w_{13}-w_{11})^2+(w_{14}-w_{12})^2)^{1/2}$ = ...

Calculate the Distance Using Xc

16a   Dist. $= |(X_1 - Xc) * w_{15}|$ =   _____

Calculate the Distance Using Yc

16b. Dist. $= |(Y_1 - Yc) * w_{15} / w_{11}|$ =   _____

Calculate the Distance Using Zc

16c. Dist. $= |(Z_1 - Zc) * w_{15} / w_{12}|$ =   _____

To check the results compute the distance using at least two of the equations above (16a through 16c).

19.2.3   Given a CS Coordinate Calculate a Distance Parallel to the YZ Plane.
Compute the values in the order given.

11. $w_{11} = w_5 / w_6$ = ...

12. $w_{12} = w_7 / w_6$ = ...

13. $w_{13} = w_8 / w_9$ = ...

14. $w_{14} = w_{10} / w_9$ = ...

15. $w_{15} = ((w_{13}-w_{11})^2+(w_{14}-w_{12})^2)^{1/2}$ = ...

Calculate the Distance Using Xc

16a   Dist. = $|(X_1 - Xc) * w_{15}|$ =   _____

Calculate the Distance Using Yc

16b. Dist. = $|(Y_1 - Yc) * w_{15} / w_{11}|$ =   _____

Calculate the Distance Using Zc

16c. Dist. = $|(Z_1 - Zc) * w_{15} / w_{12}|$ =   _____

To check the results compute the distance using at least two of the equations above (16a through 16c).

19.3 Conclusion

Many permutations of information can be used to compute object-space dimensions. Some of these permutations, not all, have been enumerated in this chapter. In all cases, at least one object-space dimension must be known to compute other object-space dimensions. In contrast to the previous chapter, this method uses camera-station coordinates as known dimensions from which other object-space dimensions are inferred.

DIMENSIONAL ANALYSIS THROUGH PERSPECTIVE

Chapter Twenty
Object-Space Coordinates

20.1 Introduction

In one-image photogrammetry it is impossible to calculate the object-space coordinates (XYZ) of a point without knowing one of the coordinates. Therefore, in the steps given in this section one of the object-space coordinates is a given known. The methods are in the order of compute X and Y given Z, compute X and Z given Y, and compute Y and Z given X. All of the Phase One and Two parametric values have been computed. In the methods presented it is assumed that the mensuration origin is not the principal point.

20.2 Compute the Object-Space Coordinates

Given one of the object-space coordinates the computations are the same until Step 6. The ID number for the point is 1. The image-space coordinates of the point are known. Compute the values in the order given.

The Initial Computations for Point 1 (E-27)

1. w_1 = $x_1 - xo$ = ...
2. w_2 = $y_1 - yo$ = ...
3. w_3 = $r_{11}{}^*w_1 + r_{21}{}^*w_2 - r_{31}{}^*f'$ = ...
4. w_4 = $r_{12}{}^*w_1 + r_{22}{}^*w_2 - r_{32}{}^*f'$ = ...
5. w_5 = $r_{13}{}^*w_1 + r_{23}{}^*w_2 - r_{33}{}^*f'$ = ...

Case One - the Known Coordinate is Z_1

6. X_1 = $Xc + ((w_3 / w_5) * (Z_1 - Zc))$ = _____
7. Y_1 = $Yc + ((w_4 / w_5) * (Z_1 - Zc))$ = _____

Case Two - the Known Coordinate is Y_1

6. X_1 = $Xc + ((w_3 / w_4) * (Y_1 - Yc))$ = _____
7. Z_1 = $Zc + ((w_5 / w_4) * (Y_1 - Yc))$ = _____

Case Three - the Known Coordinate is X_1

6. Y_1 = $Yc + ((w_4 / w_3) * (X_1 - Xc))$ = _____

7. Z_1 = $Zc + ((w_5 / w_3) * (X_1 - Xc))$ = _____

20.3 Conclusion

It should be noted that all seven equations above must be computed for each new point. Sometimes, because of our cleverness in determining the Phase One and Two parametric values, we forget the order of the equations. Also, it is wise to compute the object-space coordinates twice, if the known is a previously computed coordinate. For example, we have computed object-space coordinates for point 1 (a window corner), and we wish to compute object-space coordinates for point 2 (the other window corner). The plane of points 1 and 2 is the XZ plane, therefore points 1 and 2 should have the same Y coordinate value. Also, because the points are opposite corners of the same window, they will probably have the same Z values. In this situation, having computed the coordinates for point 1, it would be wise to compute the coordinates for point 2 using both the Y and Z coordinates. If the results do not agree, it is best to seek as much collateral support to determine the best solution to use, including the average of the repeated values.

DIMENSIONAL ANALYSIS THROUGH PERSPECTIVE

Chapter Twenty-One
Dominant Geometry Combinations of Two- and Three-Point Perspective

An important task of the photogrammetric analyst, referred to in the previous chapters, is that of review and planning. Reviewing the photographic image and planning the procedures to provide the solution to the perspective geometry is probably the single most time consuming task, and it is a task that cannot go unaccomplished. The review and plan can be as simple as deciding from the perspective geometry the procedure to use to determine:

1. Phase One parameters,
2. Phase Two parameters,
3. Model-space coordinates (object-space with scale), and
4. Required coordinates (or dimensions).

Each of these four steps is expanded into detailed step-by-step procedures referencing the solutions presented in this manual. In this chapter we present various dominant geometries of perspective photography procedures for several different possible scenarios. With these presentations and scenarios we demonstrate the use of the manual through planning statements, some detailed procedural steps, and primarily by referencing the various Phase One and Phase Two procedures already described in previous chapters.

21.1 Introduction

In single-photograph (or single-image) perspective, dominant geometry provides the best way for the photogrammetric analyst to reconstruct the three-space dimensions of the object imaged. In previous chapters, the dominant geometry was either one-, two-, or three-point perspective, and was assumed sufficient for a complete reconstruction. Now, we address the application of multiple two-point and combined two- and three-point perspective geometries. A cropped perspective image with multiple dominant geometries can require less prior information than a cropped image with single dominant geometry. Multiple perspective and combined geometries can offer enough redundancy to strengthen a weak solution. These facts motivate using multiple and combined perspective geometries in three-dimensional reconstruction.

One combined geometry scenario is multiple two-point perspective. In two-point imagery, numerous objects can rest on parallel horizontal planes, and have some of the same Phase One (image-space) parameters; e.g., vanishing points (**VP**s), principal point (**pp**), camera station (**CSXY**), effective focal length (**f'**), and rotation angles. For example, if the vanishing points for a number of buildings (in one image) lie on the same **THL**, the **CSXY**s are identical. This identity enables retrieval of the Phase One geometry required to develop the model-space coordinates, even if the imagery is cropped and no other information is available. Then, given at least one known dimension, the object-space (Phase Two) coordinates can be determined.

Another combined geometry scenario contains two-point and three-point perspective. (The geometry of three-point perspective exists in any two-point perspective image that contains an inclined plane.) When the tilt angle is 90° (and the object parallels the ground plane but is rotated to the image plane), there is two-point perspective, and vertical lines in object space are truly parallel in the image space. However, this applies only to the object that creates the two-point geometry on the image, while other objects in the image can appear as different perspective views. For example, if a Cape Cod house is photographed with a camera tilt angle of 90°, the main walls — and all planes orthogonal to the main walls — produce a dominant two-point geometry. However, the roof (an inclined plane, as illustrated in Figure 21-1) will be in a three-point perspective that shares the two-point Phase One parameters **VPY**, **pp**, and **f'**. The three-point reconstruction can be done, without full format or the parametric values of the Phase One perspective, but only after first working the two-point reconstruction. In particular the point $\mathbf{VPZ_3}$ (subscripts refer to two- or three-point perspective values) can be found graphically by constructing three perpendiculars: 1) from $\mathbf{THLXY_3}$ through the **pp**, 2) from $\mathbf{pp\text{-}VPX_3}$ through **VPY**, and 3) from **pp-VPY** through $\mathbf{VPX_3}$. The intersection of the three lines is $\mathbf{VPZ_3}$. The analytical procedure is essentially the same, making use of the two **VP**s and the **pp**. The solution to the remaining three-point Phase One parameters follows standard application procedures.

In this chapter, both analytical and graphical reconstruction applications are discussed in presenting the scenarios. (Therefore, this chapter is properly placed after the graphical and analytical parts of this manual.) Perspective reconstruction methods require that certain parametric values be either known or determined through the available geometry (Moffitt and Mikhail, 1980); our emphasis for these scenarios will be on the latter case. We also present the analytical procedures that parallel the graphical methods: conversion between two-point and three-point object-space coordinates for the same photograph, and the analytical location of the true horizon line in a two-point solution.

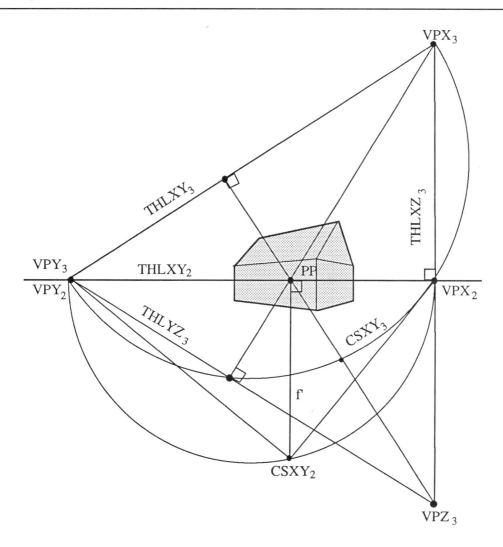

**Figure 21-1. Two- and Three-Point Perspective in
Same Image - Cape Cod House**

The dominant geometry applications presented are general in development and can be adapted to similar situations. We recommend that, prior to starting any graphical or analytical procedures, the photogrammetric analyst should include, as part of his planning list, what geometry is observed and make limited sketches of prospective procedures. Such preliminary analysis commonly reveals subtle information about the solution(s), and therefore is a good investment of time.

21.2 Analytical Solutions for Vanishing Points

The analytical methods discussed rely on some fundamental techniques to find the vanishing points. These techniques can be used no matter what the perspective in the imagery; e.g., one-, two-, or three-point perspective. For example, the image of a warehouse and loading dock illustrated in Figure 21-2 appear in two-point perspective. We use parallel lines on the horizontal planes of the warehouse and loading dock to

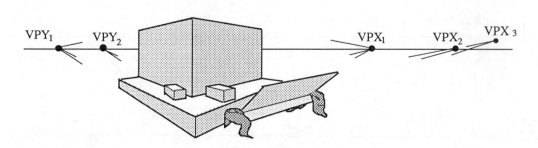

**Figure 21-2. Two- and Three-Point Perspective in
Same Image - Warehouse and Trailer**

obtain the analytical solution for a vanishing point. In this two-point perspective example there are at least eight vanishing lines converging to a vanishing point. To determine the vanishing point, one can use equations for two intersecting lines, or for three or more intersecting lines. (It is sometimes advisable to use the two-line intersection procedure for quick approximations of the vanishing-point location. The application equations for determining the intersection of two lines is given in Chapter Thirteen.) Although only two points are needed to determine a line, the solutions presented in Chapter Thirteen may not offer the desired line intersection accuracy. More measured points per line, and a least squares solution (not discussed further here), might provide more accuracy.

With only two points per line, and using multiple parallel perspective lines, a least square solution can be obtained for the best fit intersection point. It is also possible to obtain the minimum distance from each line to the computed intersection point. A program listing of this procedure is also to be found in the Appendix. The minimum distance values are used as indicators for the best solution. Any line with an exceptionally large minimum distance value can be eliminated, and the remaining data processed again, until there is an equal distribution about the intersection point.

21.3 Determining the Two-Point THL (E-28)

The examples provided in this manual have had enough sets of parallel lines to define the vanishing points; e.g., two sets of multiple lines for a two-point solution. So far we have stated that two vanishing points are needed to define a **THL**, and two sets of converging lines are sufficient to determine these vanishing points. However, this condition can often be absent in real life. For example, consider Figure 21-2b (a subimage of Figure 21-2a) that is cropped so only **VPX** can be established directly, and only the top right line of the crate is available to locate **VPY**. To determine the position of the **THL** graphically, locate **VPX**, and draw a line perpendicular to the imaged vertical lines through **VPX**: this line is the **THL**. To locate **THL** analytically use two points on any vertical line and compute the intersection point on that line of a perpendicular to the line through **VPX**. The following procedure is used to determine the coordinates of the intersection of the vertical line and the perpendicular to the vertical line through **VPX**. The points on the vertical line are numbered 1 and 2, **VPX** is point number 3, and the intersection is point number 4.

1. $dx = x_2 - x_1$ = ...
2. $dy = y_2 - y_1$ = ...
3. $w_1 = dx^2 + dy^2$ = ...
4. $w_2 = y_2 * x_1$ = ...
5. $w_3 = x_2 * y_1$ = ...
6. $w_4 = (w_3 - w_2) / w_1$ = ...
7. $w_5 = (x_3 * dx + y_3 * dy) / w_1$ = ...

8. $x_4 = w_5 * dx - w_4 * dy$ = ...
9. $y_4 = w_5 * dy + w_4 * dx$ = ...

The line defined by this new point (point 4) and the **VPX** is the **THL**. It is now possible to locate **VPY** graphically by extending the right top line of the crate to the **THL**. The analytical method to locate **VPY** is to use the two-line intersection method (paragraph 14.2), where the two lines are the **THL** and the line on the top of the crate.

21.4 Single Image with Multiple Two-Point Perspective

An example of multiple two-point perspective (see Figure 21-2) is an illustrated image of a warehouse and tilted trailer parked close to the loading dock. The loading dock contains crates of various sizes and positions, all appearing in two-point perspective. Figure 21-2, representing a cropped image frame, will be used in this discussion of multiple two-point perspective.

One of the easiest two-point solutions is when two rectangular objects, resting on parallel horizontal planes, are not parallel to each other. For example, two large crates on the loading dock rest on the same horizontal plane, but the corresponding sides of each crate are not parallel. The dominant geometry is multiple two-point perspective (see Figure 21-3). Each crate gives a different two-point perspective solution. Except for a known dimension on one of the crates, there is no additional information (such as a known diagonal angle). This is a likely problem of multiple two-point perspective.

The two crates resting on the loading dock have the same **THL**. The multiple two-point perspective of the crates will have other common parameters: **f'**, **pp**, **t**, **s**, and — most importantly — the **CS** (both graphically and analytically). In this example, to solve for the **CS** is the key to the problem. Figure 21-4 shows the vanishing points (**VPX$_1$**, **VPY$_1$**, **VPX$_2$**, and **VPY$_2$**) have been determined. The two sets of vanishing points define the diameters of two circles whose centers (C$_1$ and C$_2$) are on the **THL**. In both perspective geometries, the **CS** is defined to lie on the circle with **VPX-VPY** as a diameter. The position of **CS** is common to each of the two-point solutions, and therefore must be an intersecting point of the two circles. Two such intersection points exist, but for convenience we shall select the **CS** to be below the **THL**. The point **CS** is a vertex of the triangle C$_1$-CS-C$_2$. Two sides of the triangle are radii R$_1$ and R$_2$ of the circles, and the third side is the distance D$_{12}$ between the centers of the circles.

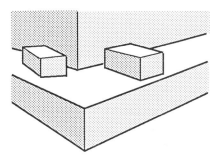

**Figure 21-3. Cropped Image, Multiple Two-Point Perspective -
Two Crates on Loading Dock**

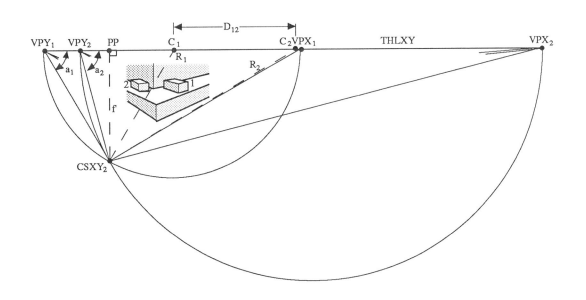

Figure 21-4. Multiple Two-Point Perspective Solution

A simple analytical procedure to determine the multiple perspective **CS** is to locate the four vanishing points **VPX$_1$**, **VPY$_1$**, **VPX$_2$**, and **VPY$_2$**. This is done for the two crates, using one of the vanishing-point procedures, and placing the origin of coordinates at **VPY$_1$** with x-axis towards **VPX$_2$**. Compute the coordinates of the centers C$_1$, C$_2$ of the two-point semicircles, and also their radii R$_1$, R$_2$. The centers are easily found by averaging the respective x and y coordinates of the vanishing points; e.g., Cx = (**VPX**x + **VPY**x)/2, and Cy = (**VPX**y + **VPY**y)/2. The radii are determined from halving the distance between the same vanishing points; e.g., R = [(**VPX**x - **VPY**x)^2 + (**VPX**y - **VPY**y)^2]$^{1/2}$/2. The distance between centers (D$_{12}$) is found in a similar manner. Compute the coordinates of the **CS** using the law of cosines on triangle C$_1$-C$_2$-**CS**. The graphical solution is shown in Figure 21-4. The analytical procedure steps are as follows:

Locate the CS Using Multiple Two-Point Perspective (E-29)

1. $w_1 = VPX_1x$ = ...

2. $w_2 = VPX_1y$ = ...

3. $w_3 = VPX_2x + VPY_2x$ = ...

4. $w_4 = VPX_2y + VPY_2y$ = ...

5. $C_1x = w_1 / 2.$ = ...

6. $C_1y = w_2 / 2.$ = ...

7. $C_2x = w_3 / 2.$ = ...

8. $C_2y = w_4 / 2.$ = ...

9. $w_5 = w_1^2 + w_2^2$ = ...

10. $R_1 = w_5^{1/2} / 2.$ = ...

11. $w_6 = (VPX_2x - VPY_2x)^2$ = ...

12. $w_7 = (VPX_2y - VPY_2y)^2$ = ...

13. $R_2 = (w_6 + w_7)^{1/2} / 2.$ = ...

14. $w_8 = (C_1x - C_2x)$ = ...

15. $w_9 = (C_1y - C_2y)$ = ...

16. $D_{12} = (w_8^2 + w_9^2)^{1/2}$ = ...

17. $w_{10} = Tan^{-1}(-w_9 / w_8) * R$ [Ang1] = ...
    (where R = 57.29578 degrees/radian)

18. $w_{11} = D_{12}^2 + R_1^2 - R_2^2$ = ...

19. $w_{12} = R_1 * D_{12}$ = ...

20. $w_{13} = w_{11} / (2. * w_{12})$ = ...

21. $w_{14} = Cos^{-1}(w_{13}) * R$ [Ang2] = ...

22. $w_{15} = w_{10} + w_{14}$ [Ang3] = ...

23. $CSx = C_1x + R_1*Cos(w_{15}/R) = CSXYx$ = ...

24. $CSy = C_1y - R_1*Sin(w_{15}/R) = CSXYy$ = ...

Once the **CSXY** is determined, the other common parameters can be determined graphically by drawing a perpendicular to the **THL** through the **CSXY**. The intersection with the **THL** is the **pp**. The line from the **pp** to the **CSXY** is the **f'**, and by definition the tilt **t** is 90°. The swing **s** is computed according to the alignment of the focal length line (principal line) and the measuring y-axis. The azimuth **a** is computed independently for each crate, and can be determined graphically as shown in Figure 21-4. The analytical solutions for these Phase One parameters are **f'** = R_1 Sin(180-Ang3) (check using **f'** = R_2 Cos(Ang1), and yo (image y of **pp**) equals the y value of the **THL**. The angles (Ang1, Ang3) are determined from the law of cosines. These equations are developed from the geometry shown in Figure 21-4. The user should be aware that the geometry depends on the position of C_1 and C_2 relative to the **pp**.

In this example, the solution of the multiple two-point perspective was developed using the fact that the two sets of vanishing points had a common **THL**. A solution is also possible with two objects imaged in a two-point perspective, but not having a common **THL**. In this case each set of **VP**s defines a different **THL**, and the intersection of the two **THL**s is the **pp**. (Note: Because the **pp** is always the same, it is impossible to have parallel **THL**s in multiple two-point perspective.) The procedures for determining the Phase One parameters are completed using each **THL** independently. The graphical solution for this example is shown in Figure 21-5.

Once all the Phase One parameters are known for the image with the two crates, it is then possible to complete the standard procedures (Phase Two) for determining the object space coordinates and dimensions (Williamson and Brill, 1987). Use the procedures presented in Chapters Eighteen through Twenty.

21.5 Combined Two- and Three-point Perspective

Consider a cropped image containing one crate on the loading dock, and a trailer next to the loading dock (Figure 21-6). The trailer is neither parallel nor perpendicular to the ground plane of the building or loading dock. On the bed of the trailer is a very large sign, centered on and coincident with the central axis plane of the trailer. The sign has three-point perspective geometry, but — as with the roof of the Cape Cod house — only two of the three vanishing points can be located. A major problem with three-point perspective is to have enough parallel lines to establish each vanishing point. Imagery with combined perspective geometry eliminates that problem, as the two-point solution provides information required by the three-point solution.

This scenario poses the problem of choosing coordinate systems for both image and object space. For simplicity of calculations it is best to transform the image coordinate system, as soon as possible, so the two-point **VPY** becomes the new origin and the **THL**$_2$ becomes the x-axis. (The subscript 2 denotes that the **THL** belongs to the two-point solution.)

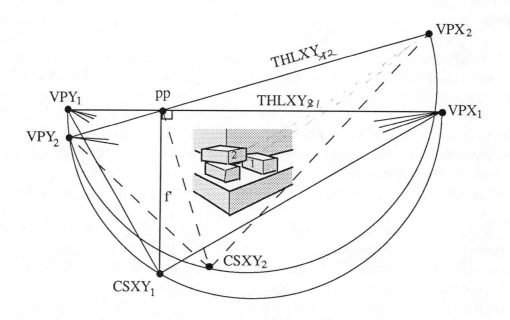

**Figure 21-5. Multiple Two-Point Perspective -
Non-parallel THLs and a Common pp**

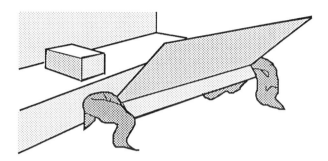

Figure 21-6. Combined Two- and Three-Point Perspective Example

Combined two- and three-point perspective calculations also require common object-space coordinates. Unlike three-point perspective, two-point perspective involves two vertical planes (XZ and YZ). Hence the object-space coordinate systems imposed by two- and three-point geometries are shifted and rotated relative to one another. Both the two- and three-point coordinate systems are orthogonal right-handed systems. Combining the systems requires using one object-space coordinate system to define point locations in both object-space systems. Ideally, this is developed using the projective equations and a point common to both geometry sets.

To convert the three-point object-space coordinates to two-point object-space coordinates, combine the two- and three-point forms of the projective equations to obtain

$$
\begin{bmatrix} X_j \\ Y_j \\ Z_j \end{bmatrix}_{2p} = \begin{bmatrix} Xc \\ Yc \\ Zc \end{bmatrix}_{2p} + [R]^T_{2p} \ [R]_{3p} \begin{bmatrix} X_j - Xc \\ Y_j - Yc \\ Z_j - Zc \end{bmatrix}_{3p} \qquad \text{(Eq. 21-1)}
$$

The procedure for using these equations is straightforward in application (Williamson and Brill, 1989). The Phase One and Phase Two parametric values for both the two- and three-point solutions have already been found, and the object-space coordinates for the point in the three-point solution are determined. This information determines the right-hand side of Eq. 21-1, and hence the XYZ coordinates in the two-point solution on the left-hand side of Eq. 21-1. The important features of this procedure are: 1) Phase One and Two parameters are computed only once, 2) the three-point XYZ values are computed for each point, and 3) the three-point XYZ values cannot be computed without knowing one of the XYZ values.

Transformation of Object-Space Coordinates

Three-Point to Two-Point (E-30)

(As stated $[M]=[R]_{2p}$, $[M]^T=[R]^T_{2p}$, and $[C]=[M]^T[R]_{3p}$)

1. $c_{11} = m_{11}r_{11} + m_{21}r_{21} + m_{31}r_{31}$ $\qquad$ = $\qquad$ ....
2. $c_{12} = m_{11}r_{12} + m_{21}r_{22} + m_{31}r_{32}$ $\qquad$ = $\qquad$ ....
3. $c_{13} = m_{11}r_{13} + m_{21}r_{23} + m_{31}r_{33}$ $\qquad$ = $\qquad$ ....
4. $c_{21} = m_{12}r_{11} + m_{22}r_{21} + m_{32}r_{31}$ $\qquad$ = $\qquad$ ....
5. $c_{22} = m_{12}r_{12} + m_{22}r_{22} + m_{32}r_{32}$ $\qquad$ = $\qquad$ ....
6. $c_{23} = m_{12}r_{13} + m_{22}r_{23} + m_{32}r_{33}$ $\qquad$ = $\qquad$ ....
7. $c_{31} = m_{13}r_{11} + m_{23}r_{21} + m_{33}r_{31}$ $\qquad$ = $\qquad$ ....
8. $c_{32} = m_{13}r_{12} + m_{23}r_{22} + m_{33}r_{32}$ $\qquad$ = $\qquad$ ....
9. $c_{33} = m_{13}r_{13} + m_{23}r_{23} + m_{33}r_{33}$ $\qquad$ = $\qquad$ ....

Steps 10 through 12 are three-point object-space coordinates

10. $w_1 = X_j - Xc$ $\qquad$ = $\qquad$ ....
11. $w_2 = Y_j - Yc$ $\qquad$ = $\qquad$ ....
12. $w_3 = Z_j - Zc$ $\qquad$ = $\qquad$ ....

Now complete the transformation from the three-point object-space coordinate system to the two-point object-space coordinate system. The results of Steps 13 through 15 are two-point object-space coordinates.

13. $X_j = Xc_{2p} + (w_1 c_{11} + w_2 c_{12} + w_3 c_{13})$ $\qquad$ = $\qquad$ _____
14. $Y_j = Yc_{2p} + (w_1 c_{21} + w_2 c_{22} + w_3 c_{23})$ $\qquad$ = $\qquad$ _____
15. $Z_j = Zc_{2p} + (w_1 c_{31} + w_2 c_{32} + w_3 c_{33})$ $\qquad$ = $\qquad$ _____

The step-by-step procedures for completing the photogrammetric analysis are presented in the following paragraphs. Although very tedious if done using a non-programmable calculator, the procedures can easily be programmed for desktop computers, and are streamlined by using a digitizer to collect the image data.

The required Phase One parameters for two- and three-point perspective analysis are those listed in Chapter Eleven. Each perspective solution will have the same basic set of parametric values, with the three-point solution having more because of the third vanishing point. The methods presented in Chapters Twelve through Sixteen are used to determine the parametric values.

Continuing the description of the procedures, assume that a known vertical distance (V) is to be used in both the two- and three-point solutions to determine the object-space camera station CS (a Phase Two computation). Because the distance V is known, the coordinates of the end points can be assigned; e.g., to (DX_1, DY_1, DZ_1) and (DX_2, DY_2, DZ_2), where $DX_1 = DX_2$, $DY_1 = DY_2$, $DZ_1 - DZ_2 = V$. Any other known dimension, in the orthogonal planes, may be substituted and the coordinates assigned accordingly. For instance, a horizontal dimension (H) in the X or Y direction could be used, and $DX_1 - DX_2 = H$ or $DY_1 - DY_2 = H$. However, when the dominant geometry of the image is two-point, it is best to use a vertical line, which is represented true-view (not foreshortened by perspective).

The equations for the CS are defined in three cases—XY, XZ, and YZ—depending on which plane contains the known dimension. Use the procedures in Chapter Eighteen to calculate the camera station coordinates. Keep in mind that because there are two different coordinate systems, with different origins and orientations (two-point vs three-point solutions, the conversion does not occur until after the solution has been completed), there will be two sets of camera station coordinates, even though in reality there is only one camera station.

21.6 Two-Point Solution in the Combined Image (E-31)

For the two-point analysis without full format, the origin of the mensuration coordinates is best selected such that mensuration values are positive. The following steps determine the Phase One values for the two-point perspective of Figure 21-6 (using only the right-hand crate), given a known vertical-diagonal angle. The application methods used have been described in this manual.

Step 1. Locate VPX_2. Using three or more parallel lines, measure two points per line, and calculate the image coordinates $VPXx_2$ and $VPYy_2$ (see Chapter Thirteen).

Step 2. Locate VPY_2. Select three or more lines, and using the same method as in Step 1, determine the coordinates $VPYx_2$ and $VPYy_2$.

Step 3. Transform the image coordinate system so VPY_2 becomes the new origin, the x-axis is defined coincident to the THL_2, and +x is towards VPX_2. The equations for this are:

1. w_1 = $VPXx_2 - VPYx_2$ = ...
2. w_2 = $VPYy_2 - VPXy_2$ = ...
3. w_3 = $x_i - VPYx_2$ = ...
4. w_4 = $y_i - VPYy_2$ = ...
5. x_j = $(w_3{*}w_1 - w_4{*}w_2)/(w_1{}^2 + w_2{}^2)^{1/2}$ = _____
6. y_j = $(w_3{*}w_2 + w_4{*}w_1)/(w_1{}^2 + w_2{}^2)^{1/2}$ = _____

Repeat Steps 3 through 6 for each transformation point. Then determine the coordinates of the midpoint **C** between VPX_2 and VPY_2 on the **THL**.

7. Cx = $VPXx_2 / 2$. = ...
8. Cy = $VPXy_2 / 2$. = ...

Step 4. Locate the vertical for **VPDVX**. Using **VPX-C** as one leg of a 45° right triangle, calculate the endpoint of the end of the leg (vertical side) of the right triangle. The vertical side will be a line through VPX_2 perpendicular to THL_2. The **VPDVX** will be located on this line.

Step 5. Locate **VPDVX**. Extend the known image diagonal of the crate until it intersects with the vertical line of Step 4. Use the two-line intersection equations to determine the intersection, which is the vanishing point **VPDVX** (with image coordinates VPDVXx, VPDVXy). If there are several known diagonal angles (1 through k), then there can be as many $VPDVX_k$ positions. Each position should be determined, which in turn will help determine the location of the measuring point (or camera station revolved) **MPX**.

Step 6. Locate **MPX** (**CSXY** rotated). With the position of **VPDVX** and the true angle (TA) of the diagonal known, calculate the **MPX** coordinates using MPXx = VPXx - (VPDVXy - VPXy) Tan(TA/Rad), and MPXy = 0.

Step 7. Locate **CSXY**. The graphical location of **CSXY** involves rotating the point **MPX**, with **VPX** as the center of rotation, to intersect with the semicircle for **VPX-VPY**. Analytically, the solution involves using the isosceles triangle **CSXY-C-VPX** with the radius of the semicircle as the two equal sides and the midpoint of **VPX-VPY** as the apex. The side opposite **C** is **VPX-CSXY**, which is equal to the line **VPX-MPX**. The angle (Ang1) at **VPX** is determined using the law of cosines and the sides of the isosceles triangle. The cosine of Ang1 is computed using cos(Ang1) = (**VPX**x-**MPX**x)/2(**VPX**x-Cx). The sine of Ang1 of found using sin(Ang1) = $(1.0-\cos(\text{Ang1})^2)^{1/2}$. The **CSXY** coordinates are computed using CSXYx = VPXx - (VPXx-MPXx)Cos(Ang1), and CSXYy = -(VPXx-MPXx)Sin(Ang1).

Step 8. Determine remaining Phase One parameters. The **pp**, **f'**, **a**, and **s** can now be determined from the image coordinates of **VPX**, **VPY**, and **CSXY**. The tilt angle, by definition is 90°. The swing angle will be 180°, if the image coordinate axis for x is parallel to both the **THL** and the bottom edge of the frame. The angle **a** is found using right-triangle **VPY-THxy-CSXY**, and the equation is $a = \tan^{-1}((\text{CSXYx-VPYx})/\text{CSXYy})$. Equations to determine the rotation matrix are presented in Chapter Fourteen.

Step 9. Locate **XC, YC, ZC**. Given the two-point perspective Phase One parameters, determine the Phase Two camera-station coordinates. Use the procedures presented in Chapter Eighteen. First compute one **CS** coordinate, and then using that value compute the remaining two **CS** coordinates. For two-point perspective, the value of VPZx will equal **pp**x, and the value of VPZy should be an extremely large number; e.g. $|1 \times 10^{16}|$. It is recommended that you use these values of **VPZ** only as a means to check the results (if possible). The value of **t** is 90°. It is also possible to verify the results with a graphical solution for the **CS**.

Step 10. Determine object-space coordinates. Use the procedures presented in Chapter Twenty to determine the two-point system object-space coordinates. Remember, for each image point you will need the Phase One and Phase Two values, the image coordinates, and one object-space coordinate. This is a building block procedure.

21.7 Three-Point Solution in the Combined Image (E-32)

The procedures for the three-point solution of Figure 21-6 are a continuation of the two-point procedures.

Step 11. Locate VPX_3. Compute the image coordinates of VPX_3 using the same procedures described in Step 1.

Step 12. Locate VPZ_3. Using the same procedures of Step 1, determine the image coordinates of VPZ_3.

Step 13. Locate VPY_3. With the known image coordinates of the **pp**, VPX_3, and VPZ_3, compute VPY_3 using the equations of Procedure No. 3, paragraph 14.6.

Step 14. Locate the f'_3. Determine the effective focal length for the three-point solution by using the procedures presented in Chapter Fifteen.

Step 15. Compute **[R]**. Calculate the rotation matrix elements using the procedures presented in Chapter Fourteen.

Step 16. Locate **XC, YC, ZC**. Calculate the camera station coordinates for the three-point geometry using the same procedures used in Step 9 (Chapter Eighteen).

Step 17. Locate object-space points. Use the same procedures as in Step 10 to locate the three-point solutions object-space coordinates.

Step 18. Convert the object-space coordinates using the procedures presented earlier in this chapter.

These 18 steps complete the analytical procedures for this scenario of combined two- and three-point perspective. There was a mixture of analytical and graphical procedures where the method is best shown or described graphical and completed analytically. The results of the graphical procedures are shown in Figure 21-7.

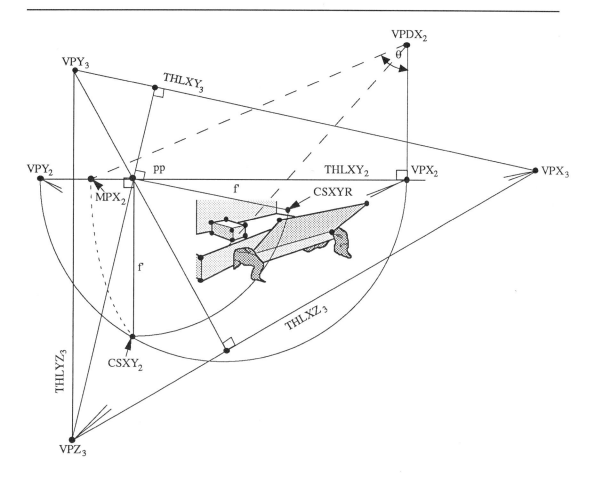

Figure 21-7. Combined Two- and Three-Point Perspective Solution

21.8 Conclusion

Although three-dimensional reconstruction from a single image always requires some prior information about object space, less information is needed when there is multiple or combined dominant geometry than when the geometry is simple. To illustrate this fact, we chose a multiple two-point perspective in which the imaged objects rest on a common horizontal plane. In cropped imagery with simple two-point perspective, a horizontal or vertical diagonal angle had to be known to complete the reconstruction (Williamson and Brill, 1987). However, such an angle need not be known for cropped imagery with multiple two-point perspective in which nonparallel objects share the same ground plane or share parallel ground planes. Under these conditions the camera station is found by locating the vanishing points for each nonparallel object, constructing the circles defined by the respective sets of vanishing points, and identifying an intersection point of circles as the camera station. With three or more sets of vanishing points, the precision of the method is indicated by how close the circles come to intersecting at a point. If the respective object horizontal planes were not parallel, but each offered two-point perspective, the camera station could emerge with equal ease by noting that the unique principal point is the intersection between the two true horizon lines (**THLs**).

Nonparallel ground planes that are perpendicular to the image plane are rare. More frequent are planes that are skewed both to the two-point horizontal plane and to the image plane (as in three-point perspective). However, as was the case with the roof of the Cape Cod house and the sign on the trailer, such planes are often not attached to deep enough rectangular structures to enable determination of more than two vanishing points.

It has been shown that available two-point geometry can be used to augment the three-point problem to retrieve the missing three-point vanishing point. As in most combined-perspective geometry problems, there is a need to embed the solutions in a single object-space coordinate system. This can be done analytically using a dimension of known length parallel to one of the major orthogonal coordinate planes.

From the scenarios presented, it should be clear that combined perspective geometries in a single image can serve not only to fill in missing information, but also to strengthen geometry by using redundant information to enhance the precision of the solution. In this chapter, emphasis was given to some analytical representations, with recommendations that the procedures be placed on a programmable calculator or desktop computer system for use in a close-range photogrammetric workstation. Such a workstation, developed for multiple close-range digital imagery (Williamson, et al, 1988), incorporates these single-image perspective procedures as a secondary system.

DIMENSIONAL ANALYSIS THROUGH PERSPECTIVE

Chapter Twenty-Two
Summary

This manual has presented our concepts of various methods of single-image perspective analysis through a series of specific chapters. Chapters One and Two provide the analyst information about this manual. In Chapters Three through Seven, the stages of the graphical methods are introduced starting with simple constructions (demonstrating how to locate vanishing points and the Phase One parameters), and then extended to complicated stages of dimensional analysis of elementary objects seen in one-, two-, and three-point perspective. There are worked examples in the Appendix that demonstrate the procedures described. Chapters Nine through Twenty provide the stages for the analytical methods corresponding to the graphical methods, and were developed following the same progression from simple beginnings to a complicated but elementary analysis. The Appendix also includes worked examples to support the procedures described. In each stage of analysis, we used a step-by-step development, both in the detail and in organization, and emphasized this development in the organization of the manual.

Following these chapters we introduce in Chapter Twenty-One the development of interrelated single-image geometries inhabiting the same image. The task confronting the analyst in this Chapter is more complicated than in previous chapters, for it requires the integration of partial solutions from several dominant geometries to complete the perspective analysis and obtain model- or object-space values.

By now it should be apparent that mensuration of true-life, complicated images requires logic, cleverness, and <u>practice</u>. It is not always immediately obvious how to extend the building blocks (or chain of geometric inference) from a known dimension to an unknown dimension. There will be times when the extension seems impossible (and in fact may be impossible), and one must settle for an approximation to the dimensions. This does not invalidate the geometry of the photograph, it only develops a model that does not match the scale of object-space. Since man-made objects are usually predominate in the image, there should be enough information about angles and geometric shapes to obtain a relative solution. (The relative solution produces reconstruction of model-space, as we have mentioned in several places—especially in Chapter Seven.)

It is possible that the geometries of an image are so undecipherable that the image cannot be used to provide perspective analysis. However, for most applications of concern to photogrammetric analysts, much progress can be made in providing relative reconstruction (or model reconstruction) of objects within an image, and in some fortunate cases the object-space reconstruction from this model can be determined from a single object-space dimension. We hope this manual has provided enough examples and explanations to encourage the photogrammetric analyst to venture forth into the terra incognita of perspective images, and to bring them under control to the greatest extent possible.

APPENDIX

A-1: Mini-Log for Camera Collections

Date/time of collection: _____ / _____ / _____ : _____ A/P _____

Camera type/name: _____

Camera serial number: _____

Lens serial number: _____

Nominal focal length: _____

Range/known dimensions: _____

_____

Target: _____

Camera station ID: _____

Weather conditions: _____

Operator height:_____

WAS LOG FILLED IN BEFORE OR AFTER COLLECTION?: _____

A-2: Intersection of Multiple Lines, with Minimum Distance

The following program source code is a routine to simultaneously determine the intersection of multiple lines. The source code establishes the matrix of coefficients for a Gaussian adjustment, the intersection point is determined, and then the minimum distance from each line to the intersection point. The minimum distance is a method of determining if a line is acceptable to the solution. In this code the terms X(N), Y(N), X(NP1), and Y(NP1) are the coordinates of two points on a line. The number of lines read is N/2. The term D(K) is the minimum distance of each line K to the intersection point XI, YI, and the number of ~~lines~~ *Points* is NMAX.

Multiple Line Intersection

```
    DO  10  N  =  1,NMAX,2
      NP1 = N + 1
      READ * X(N),Y(N),X(NP1),Y(NP1)
      Q(1) = Y(NP1)-Y(N)
      Q(2) = X(N) - X(NP1)
      Q(3) = Y(N)*X(NP1) - X(N)*Y(NP1)  * -1
      DO  10  I  =  1,2
        DO  10  J  =  1,3
        C(I,J) = C(I,J) + Q(I)*Q(J)
10    CONTINUE
      CALL [GAUSSIAN ROUTINE](C,2)
      XI = C(1,3)
      YI = C(2,3)
      WRITE * XI,YI
      K = 0
      DO  20  M  =  1,NMAX,2
        K = K + 1
        MP1 = M + 1
        Q(1) = Y(MP1) - Y(M)
        Q(2) = X(M) - X(MP1)
        Q(3) = Y(M)*X(MP1) - X(M)*Y(MP1)
        S = SQRT(Q(1)**2 + Q(2)**2)
```

```
     D(K) = (Q(1)*XI + Q(2)*YI + Q(3))/S
     WRITE * D(K)
20   CONTINUE
```

Gaussian Routine

Gaussian solution of N unknowns and M equations. The M equations have already been normalized and are in [A]. This is a double-precision routine.

```
     DIMENSION A(4,4)
     NP1 = N + 1
     NM1 = N - 1
     DO 1 K = 1,NM1
      KP1 = K + 1
      DO 2 J = KP1,NP1
      A(K,J) = A(K,J) / A(K,K)
      2 CONTINUE
      DO 3 I = KP1,N
       DO 3 J = KP1,NP1
       A(I,J) = A(I,J) - A(I,K)*A(K,J)
3    CONTINUE
1    CONTINUE
     A(N,NP1) = A(N,NP1)/A(N,N)
     DO 4 M = 1,NM1
      I = N - M
      IP1 = I + 1
      DO 5 K = IP1,N
       A(I,NP1) = A(I,NP1) - A(I,K)*A(K,NP1)
5    CONTINUE
4    CONTINUE
```

A-3: List of References

Beamish, J. (Editor - 1984). <u>Close-Range Photogrammetry & Surveying: State-of-the-Art</u>, Proceedings of a Workshop, American Society of Photogrammetry, Falls Church, Virginia.

Brill, M. H. and J. R. Williamson (1987), "Three-Dimensional Reconstruction from Three-Point Perspective Imagery," <u>Photogrammetric Engineering and Remote Sensing</u>, Vol 53 , 1679-1683.

Busby, W. E. (1981), "Dimensional Analysis of a Single Terrestrial Photograph Using the Principles of Perspective," <u>Proceedings of the Fall Symposium</u>, The American Society of Photogrammetry, pp. 431-446.

Ethrog, U., 1984, "Non-Metric Camera Calibration and Photo Orientation using Parallel and Perpendicular Lines," <u>Photogrammetria</u>, Vol. 39, No. 1, pp. 13-22.

French, Thomas E. (1957), <u>Engineering Drawing</u>, A Manual of Engineering Drawing for Students and Draftsmen, McGraw-Hill Book Company, Inc., New York.

Gracie, G., et al. (1967). <u>Mensuration Procedures in Terrestrial Photography, Vol 2, Handbook of Analytical Procedures</u>, Naval Air Systems Command, Washington. D. C.

Kelley, C. P. (Seminar Notes). (1978-1983). <u>Some Aspects of Graphical Techniques in the Rectification of Hand-Held Single Frame Photography</u>, NDHQ, Ottawa

McCartney, T. O., (1963), <u>Precision Perspective Drawing</u>, McGraw-Hill Book Company, New York.

Moffitt, F. H. and E. M. Mikhail, <u>Photogrammetry</u>, 3rd Ed., Harper and Row, New York, 1980.

Perfect, Hazel, 1986, "How obvious is it?", **<u>Mathematical Spectrum</u>**, Vol. 19, pp. 7-8.

Slama, C. C. (Editor, 1980), <u>Manual of Photogrammetry, Fourth Edition</u>, American Society of Photogrammetry, Falls Church, Virginia.

Walters, Nigel V., and John Bromham, (1970). <u>**Principles of Perspective**</u>, Whitney Library of Design, an imprint of Watson-Guptill Publication, New York.

Williamson, J. R. and M. H. Brill (1987), "Three-Dimensional Reconstruction from Two-Point Perspective Imagery," <u>Photogrammetric Engineering and Remote Sensing</u>, Vol 53, pp. 331-335.

Williamson, J. R., M. H. Brill, D. McCoy, J. T. Yezek, and B. L. Schrock, January 1988, "Close-Range Video-Imaging Photogrammetry (CRVIP) System," Final Report, SAIC Internal Research and Development Project 88/910.

Williamson, J. R. and M. H. Brill (1989), "Dominant Geometry Combinations of Two- and Three-Point Perspective in Close-Range Applications," <u>Photogrammetric Engineering and Remote Sensing,</u> Vol 55, pp. 223-230.

Williamson, J. R. and M. H. Brill (1989), "Close-Range Exploitation Workstation," Proceeding of the ASPRS/ACSM Annual Convention, April 2-7 1989, Baltimore, MD, Vol-1, pp 54-63

Wolf, Paul R., 1974, <u>**Elements of Photogrammetry**</u>, McGraw-Hill, Inc., New York, New York

Yacoumelos, N. G. (1970). "Reverse Photogrammetric Procedure in Architectural Design," Paper Presented to the Fall Technical Conference at the Denver ASP-ACSM Convention, October 7-10.

A-4: About the Authors

Dr. James R. Williamson received his B.S. in Civil Engineering at the University of Texas, 1965, an M.S. in Civil Engineering at the University of Illinois, 1972, and a Ph.D. in Engineering from Kennedy Western University, 1986. He has organized and taught many short courses on single-photo analytical and graphical perspective, developing several innovative photogrammetric methods and algorithms. He has been a member of the American Society of Photogrammetry and Remote Sensing since 1966. Dr. Williamson has made significant contributions to many areas of photogrammetry for over 30 years in private practice, industry and government. As a Senior Staff Scientist for Science Applications International Corporation, Falls Church, VA, Dr. Williamson is managing the development of a SAIC Close-Range Exploitation Workstation (SAIC CREW), and other projects involving photogrammetry.

Dr. Michael H. Brill received his B.A. in Physics at Case Western Reserve University, 1969, an M.S. in Physics at Syracuse University, 1971, and a Ph.D. in Physics at Syracuse University, 1976. He has been a member of the American Society of Photogrammetry and Remote Sensing since 1985. He has given many invited lectures and seminars on computational vision and photogrammetry. An internationally recognized authority in computational vision, Dr. Brill has authored or co-authored over 40 refereed articles in the mathematics of visual perception (especially color), computer vision, photogrammetry, and projective geometry. Dr. Brill is a Senior Staff Scientist for Science Applications International Corporation, McLean, VA, where he develops innovative solutions to geometrical problems in optics, acoustics, photogrammetry, and image understanding.

A-5: WORKED EXAMPLES Example Problems List

Graphical Examples:

E-1: One-Point Perspective of Hall
E-2: One-Point Perspective of Hall with Open Door and Room at End of Hall.
E-3: Airport Hangar with Cart
E-4: Rectangular Solid - Two-Point Perspective, with Alternatives
E-5: Rectangular Solid - Three-Point Perspective
E-6: Non-Rectangular Solid - Three-Point Perspective of Pentagon

It is understood that the paragraphs relating to these exercises have been read, eliminating the need to repeat detailed descriptions here.

Recommended basic items (equipment):

1. A clear acetate overlay
2. A pin-prick (needle point)
3. Colored pencils
4. Facial tissues
5. Triangles (45° and 60°)
6. An engineering scale
7. Protractor (180° or 360°)
8. Dividers (needle point)
9. Pocket calculator (scientific)

To draw a graphical line means to use the pin-prick and scribe (scratch) a line on the acetate. Fill in line by rubbing with a colored pencil, and wipe away excess. Save time and effort by using pre-scribed strips of acetate to locate VPs, and other points where lines obscure image (see Paragraph 3.2). Proportional dividers are used as optional equipment. All points are labeled as they are constructed, or you are asking for Murphy's help.

Example E-1: One-Point Perspective of Hall

Purpose is to use known dimension to find **SF** in plane ABCD of one-point perspective image. Observations are full-format image (Fig. E-1), from a camera pointed down the center of the hall. Image plane and end wall (ABCD) are parallel (**t** = 90°). The door in the end wall and the right wall door are identical. Paragraph 5.2 gives additional information. The step-by-step solution is:

<u>Required Information</u>: Known dimension is line LM (KD_{LM})

Step 1: Draw (extend) the format side lines to corners, C1, C2, C3, and C4 (corner intersection lines only).

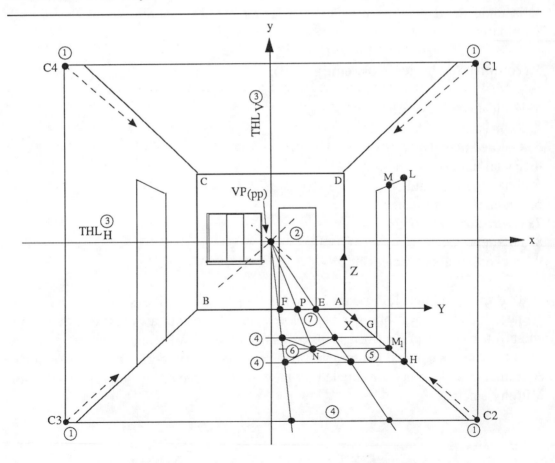

Figure E-1. One-Point Perspective of a Hall

Step 2: Draw intersection of C1C3 and C2C4 at the center-of-format (CoF). The CoF represents the **pp**.

Step 3: Draw the THL_H parallel to AB and CD (horizontal lines) through the **pp**. Then draw the THL_V perpendicular to THL_H, through the **pp**.

Step 4: Extend lines **pp**E and **pp**F to frame edge (C2-C3). Draw lines through G and H, parallel to THL_H, intersecting with **pp**E and **pp**F extended. This forms a one-point perspective rectangle (square: doors are identical).

Step 5: Draw parallel line to LH through M intersecting GH at M1. Then, draw a line through M1 parallel to THLH to intersect with **pp**E and **pp**F.

Step 6: Draw the diagonals of perspective square, label intersection N. They intersect on line through M1, proving known is half the door width. (Could have determined relationship of known to rectangle using proportional division methods described in paragraph 5.1.3.)

Step 7: Draw line **pp**N, intersection with EF is P. Scale-factor (SF_{ew}) for end wall is known divided by measured value of EP, or FP; e.g., $SF_{ew} = KD_{LM}/EP$.

Example E-2: One-Point Perspective of Hall with an Open Door to Room at End of Hall

Purpose is to use known dimension to find the distance (depth) between the two parallel planes (end wall of hall and end wall of room) that are parallel to the one-point perspective image plane. Also, determine $\mathbf{f'}$ for this example. Observations are the same as in E-1, use Figure E-2. Paragraphs 5.1.1 and 5.2 give additional information. The step-by-step solution is:

<u>Required Information</u>: Known dimension is line LM. Complete Step 1 through Step 7 of E-1.

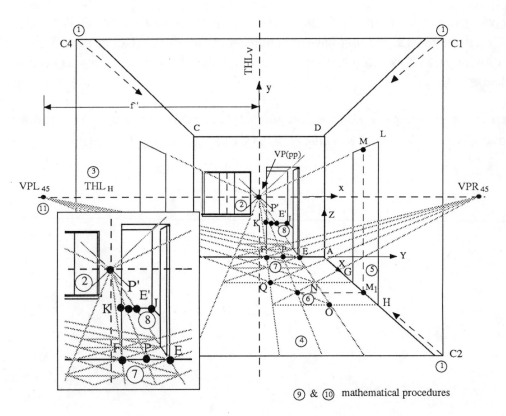

Figure E-2. One-Point Perspective of a Hall and Open Door.

Step 8: Draw lines **pp**E and **pp**P. Label the intersection points of lines **pp**E and **pp**P with line JK as E' and P' respectively.

Step 9: Scale-factor (SFrm) for room wall is the known (LM) divided by measured value of E'P', e.g., SFrw = LM/E'P'.

Step 10: Use the equation $R_{ew} = (SF_{rw} \, dR)/(SF_{ew} - SF_{rw})$, where R_{ew} is the end wall range, and dR is the distance between the two walls of known scale. The depth of the room is $dR = R_{ew} (SF_{ew} - SF_{rm}) / SF_{rm}$.

Step 11: Extend the line OQ (perspective square diagonal) to meet the THL_{H}. The intersection point is the VPL_{45}, and with the **pp** defines the radius of the base of a right circular cone. The cone is defined by the lens of the camera, and the cross-section right triangle is a 45°triangle. Thus, **f'** is the distance between **pp** and VPL_{45}. Check by using the other diagonal of the perspective square. The procedure is unique to one-point perspective and squares, and represents the diagonal azimuth angle defined as the half cone angle.

Example E-3: Airport Hangar with Cart

Purpose is to determine the length, width, and height of the crate shown in Figure E-3a. Observations are a cropped one-point perspective photograph of an airplane hangar, and hangar front and back doors are open. Hangar door opening forms

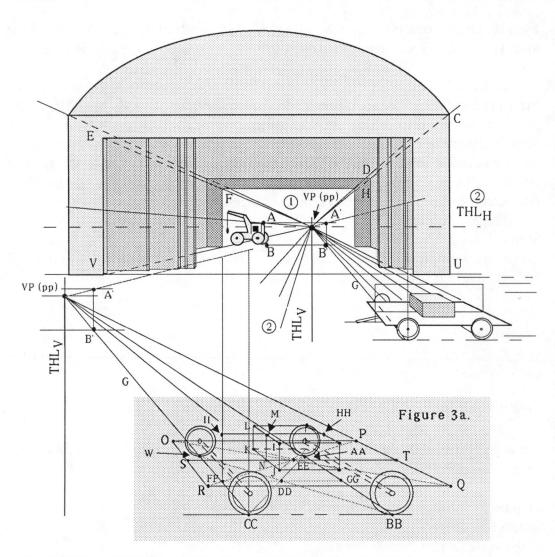

Figure E-3. One-Point Perspective of Airport Hangar with Cart

rectangle (one-point perspective, and $t = 90°$). Hangar floor and outside areas (apron) are same horizontal plane, and cart and crate are parallel to hangar. The known dimension is the height of a utility crane (apron to top of cab). Paragraphs 5.1.1 and 5.2 contain additional information.

Required Information: Known dimension is line AB.

Step 1: Locate **VP** using lines of the hangar and cart; i.e., CD, EF, PQ and so on.

Step 2: Draw lines through the **pp** that are parallel to the open hangar rectangle, label THL_H, and THL_V.

Step 3: Transfer known distance to area of interest. Draw lines parallel to THLH through A and B. Draw line **pp**CC. B' is intersection of **pp**CC with horizontal line through B. Draw a line parallel to THL_V through B' to intersect at A' (horizontal through A). Draw line **pp**A', and extend line to pass over point CC.

Step 4: Draw diagonals of left end of crate (IK, JL), and locate vertical center line (MN) (see insert from Figure E-3). Draw the diagonals of the cart top (OQ, PR) and locate horizontal center line (ST) parallel to hangar front. Do the same for the cart wheels where they touch the apron plane (WBB, AACC). The lines MN and ST intersect at N. (If they hadn't we could have used the more complicated proportional methods to locate the intersection point.) The lines ST and FFGG are in the same vertical plane, and parallel to front plane of hangar. This vertical plane also intersects with the extended plane of A'B' from Step 3, and FF is in both planes.

Step 5: Draw a line parallel to THLv through FF until it intersects with II at the extension of line HHM. This is the extension of the known line AB to the vertical center-line plane of the cart and crate. Determine the scale factor(**SFcc**) for this vertical center-line plane.

Step 6: Determine the height of the crate using **SFcc**, and then determine the scale factors SF_{IL} and SF_{JK}. Using SF_{JK} determine the length of the crate. Then using both of the scale factors determine the depth of the crate using the procedure outlined in paragraph 5.1.1.

Example E-4: Rectangular Solid - Two-Point Perspective, with Alternatives

The purpose is to develop the scaled plan and elevation views of the imaged object. Observations are that the image is a cropped two-point perspective with three known diagonal angles. Additional information is in paragraphs 6.2.2, 6.2.3, 6.2.4, and 6.2.5.

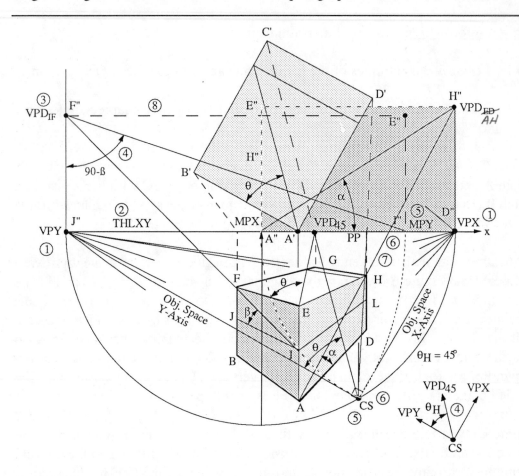

Figure E-4. Rectangular Solid - Two-Point Perspective

<u>Required values:</u> Angles, alpha (**A**) - the right face diagonal, beta (**B**) - the diagonal of half the left face, and theta (**T**) - the diagonal of the top horizontal plane. Known dimension AB.

Develop the Plan View.

Step 1: Locate **VPX** using lines AD, IL, EH, and FG. Locate **VPY** using lines AB, IJ, EF, and HG.

Step 2: Draw the **THL**, and a the semicircle with **THL** as diameter.

Step 3: Locate intersection of **THL** and diagonal line EG - this is ~~VPDH~~$_A$. VPD_{45}.

Step 4: On a clear piece of overlay scribe two lines at right angles representing **CS** and line to **VPX** and **VPY**. Between these lines draw a line at angle **A**, relative to correct VP. Each line should extend beyond its respective **VP** when acetate is placed on the semicircle.

Step 5: Locate **CS** by placing apex of acetate lines on semicircle and adjusting until all three lines pass through the respective **VPs**.

Step 6: Draw line perpendicular to **THL** through **CS**. Intersection of the line and **THL** is **pp**, and **f'** is distance between **pp** and **CS**.

Step 7: Extend all of the vertical lines (BF, AE, CG, and DH) to the **THL**, label intersection points with prime letters (A', and so on). Then project lines from **CS** through the prime points (CSB', and so on). Do not project through A'. Let the vertical reference plane and the **THL** be coincident, and A' = A".

Step 8: Draw a line parallel to **CSVPX** through A'. Draw a line parallel to **CSVPY** through A'. These are the X and Y axis lines in the model-space plan view. The intersection the X-axis and CSD' is D", and the Y-axis and CSB' is B". Check your drafting by locating C". Perpendicular lines to the X and Y axes at B" and D" should intersect the CSC' line at the same point. If it does not you will need to check all lines starting with Step 1.

Develop the Left-Face Elevation View.

Step 9: Extend line IF to intersect with perpendicular line through **VPY**. This is **VPDY**$_{IF}$. Using your protractor at **VPDY**$_{IF}$ mark a point U at the angle 90-**B** away from **VPYVPDY**$_{IF}$. Extend the line **VPDY**$_{IF}$U until it intersects with **THL**, and label the intersection point **MPY**. This is also point I".

Step 10: At **VPDY** draw a line parallel to **THL** to intersect perpendicular line from I" on **THL**. The intersection point is E". The rectangle formed by I", **VPY**, **VPDY**$_{IF}$, and E" is the model-space elevation area of IJFE. This elevation view is not at the same scale as the plan view. To obtain the same scale as the model-space plan, multiply all elevation lines by the ratio value of A'B"/I"J".

Develop the Right-Face Elevation View.

Step 11: Extend line IH to intersect with perpendicular line through **VPX**. This is **VPDX**$_{AH}$. Using your protractor at **VPDX**$_{AH}$ mark a point U at the angle 90-**A** away from **VPXVPDX**$_{AH}$. Extend the line **VPDX**$_{AH}$U until it intersects with **THL**, and label the intersection point **MPX**. This is also point B".

Step 12: At ~~**VPDX**~~ [VPD$_{AH}$] draw a line parallel to **THL** to intersect perpendicular line from ~~**B"**~~ [E'] on **THL**. The intersection point is E". The rectangle formed by ~~**B"**~~ [E], ~~**VPY**~~ [VPX], **VPD**$_{AH, 45}$, and E" is the model-space elevation area of AEHD. This elevation view is not at the same scale as the plan view. To obtain the same scale as the model-space plan, multiply all elevation lines by the ratio value of (2*~~A~~'E'/A"E"), [I'E'] where ~~A'E'~~ is from the left-face elevation after it has been scaled to match the plan view.

Example E-5: Rectangular Solid - Three-Point Perspective

The purpose is to develop plan and elevation views of the imaged object, at some arbitrary model-space scale. Observations are that the image is a cropped three-point perspective with no given information. Paragraph 7.2 contains additional information.

Required values: None

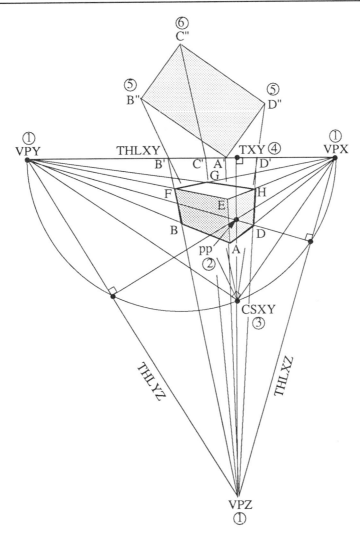

Figure E-5. Rectangular Solid - Three-Point Perspective

Step l: Extend three sets of three parallel lines so they intersect at the respective **VP**s; e.g. **VPX:** AD, EH, FG; **VPY:** AB, EF, HG; and, **VPZ:** FB, EA, and HD.

Step 2: Locate **pp** as the intersection of the three altitudes of triangle **VPX-VPY-VPZ**. (Perpendicular line to each **THL** through the opposite **VP**.)

Step 3: Construct a semicircle whose diameter is a **THLXY**. (Since there are three **THL**s, there are three semicircles.) The semicircle should be drawn so it lies inside the triangle of **VP**s. The intersection of the semicircle, and the altitude drawn through VPZ, is the **CSXY**.

Step 4: Extend lines from **VPZ** through A, B, C, and D that intersect with **THL**, and label intersection points A', B', C', and D'.

Step 5: Construct a line through A' parallel to **CSXY- VPY**, and extend **CSXY**-D' to intersect the line at point D". Then construct a line through A' parallel to **CSXY-VPX**, and extend **CSXY**-B' to intersect the line at point B".

Step 6: Construct C" by completing the rectangle A'B"C"D". (A test of the drawing accuracy is the proximity of C" to the extended line **CSXY**-C'). If the intersection at C" is poor, check your procedures starting with Step 1. This scale for this model-space view is independent of the other model-space views.

Step 7: To develop the left-face elevation view start with Step 1 and repeat the same procedures using **THLYZ** and **CSYZ**. Be sure to watch how you label the various intersection points. The scale for this model-space view is independent of the other model-space views.

Step 8: To develop the right-face elevation view start with Step 1 and repeat the same procedures using **THLXZ** and **CSXZ**. Be sure to watch how you label the various intersection points. The scale for this model-space view is independent of the other model-space views.

Example E-6: Non-Rectangular Solid - Three-Point Perspective of Pentagon

The purpose is to develop a plan view of an object that is not orthogonal, yet is a man-made geometric structure (see Figure E-6a). The image of the pentagon (E-6b) is an example of such a structure. Observations are that the image is a cropped three-point perspective image. The geometric shape of the object is known, and no knowns are provided. Paragraph 7.5.1 provides more information.

Required values: Geometric shape of pentagon, therefore the interior angle at any given vertices is 108°. Let this be angle **p**.

Step 1: Locate four vanishing points by extending the lines along four of the sides of the pentagon (12, 23, 54, and 15). Use parallel lines from the tops and bottoms of

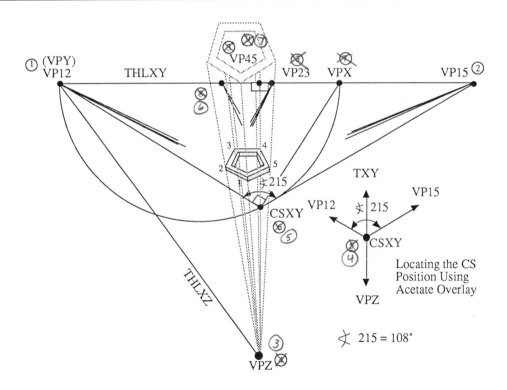

Figure E-6a. Non-Rectangular Solid - Three-Point Perspective Image of the Pentagon

windows, the parapet lines, and so on (which are omitted from Figure E-6a to simplify the illustration), and **VPY = VP12**.

Step 2: A test of your drafting accuracy is that the vanishing points **VP12**, **VP45**, **VP23**, and **VP15** should lie on the same line. If not then repeat Step 1 to relocate those in error.

Step 3: Locate **VPZ** using as many vertical lines as possible. Now you have **VPY**, **VPZ**, and **THLXY**, and can draw the altitude line from **THLXY** to **VPZ**.

Step 4: On a piece of clear plastic, draw two lines intersecting at angle **P**, and place the intersection point on the line **VPZ-TXY**. Adjust the position of the vertex until the lines pass through **VPY** and **VP15**.   This locates the **CSXY**.

Step 5: Draw a perpendicular to **VPY-CSXY** through **CSXY** to intersect with **THLXY**. The intersection point is **VPX**.

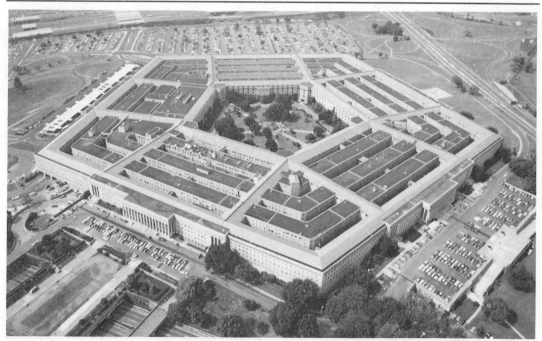

Figure E-6b. Three-Point Perspective Photo of Pentagon

(Actually you don't need the **VPX** to develop the plan view, but we thought you needed the exercise. It does help to develop the Phase One and Phase Two analytical values.)

Step 6: Project all of the vertices to intersecting point on the **THLXY**, and label them with <u>prime letters</u>; e.g., 1 = A', 2 = B', 3 = C', 4 = D', and 5 = E'.

Step 7: Project lines from the **CSXY** through the prime letters. At point A', on the **THL** construct lines parallel to **CSXY-VP15**, and **CSXY-VPY**. Where these lines intersect **CSXY**-E' and **CSXY**-B', respectively, are the pentagon corners E" and B".

Step 8: Use your protractor and construct a line 108° from line A'E", at E", construct a line 108° from line A'B" at B". These two lines should intersect **CSXY**-D" and **CSXY**-C" respectively.

Step 9: Test your results. The angles B"C"D and C"D"E" should equal 108°, and if they do not you need to start over again.

A-6: WORKED EXAMPLES
Example Problems List

Analytical Examples:

Phase One Parameters

E-7 & E-13: Center-of-Format - Intersecting Lines
E-8:   Center-of-Format - Two Points per Side
E-9:   Find **pp** Using **DKA** in Horizontal Plane
E-10:   Find **pp** Using **DKA** in Vertical Plane - **VPY**
E-11:   Find **pp** Using **DKA** in Vertical Plane - **VPX**
E-12:   Find **pp** or Major **VPs** - Orthocenter Method
E-14:   Find **VP** - Segmented Line Method
E-15:   Find **VP** - Orthogonal Line Method- (two)
E-16:   Find **VP** - Use Interior Parameters and [**R**]
E-17:   Find [**R**] Using **a**, **t**, and **s**
E-18:   Find [**R**] Using **pp**, **f'**, **a**, and **VP**- (three)
E-19:   Find [**R**] Using **pp**, **f'**, and **VPX**, **VPY**, **VPZ**
E-20:   Find **f'** Using **pp** and Two **VPs**- (three)
E-21:   Find **a**, **t**, and **s** Using [**R**]
E-22:   Find **a**, **t**, and **s** Using **pp**, **f'**, Major **VPs**

Phase Two Parameters

E-23:   Find One **CS** Coordinate (**Xc**, **Yc**, or **Zc**)
E-24:   Find Two **CS** Coordinates (**XcYc**, **XcZc**, or **YcZc**)
E-25:   Find Three **CS** Coordinates (**Xc**, **Yc**, and **Zc**)
E-26:   Find Object-Space Dimension
E-27:   Find Object-Space Coordinates
E-28:   Find the Two-Point **THL**
E-29:   Find **CS** Using Multiple Two-Point Perspective
E-30:   Transformation from Three-Point into Two-Point Object-Space
                  Coordinates
E-31:   Two-Point Solution in Combined Geometry Imagery
E-32:   Three-Point Solution in the Combined Geometry Imagery

Example E-7 & E-13: Center-of-Format - Intersecting Lines

The purpose of this exercise is to determine line intersections; e.g., center-of-format using fiducial points (f_1 - f_4), a line intersecting the **THL**. The center-of-format calculations are shown. (Test yourself, solve for line AC intersecting **THL** at **VPDH**.) Refer to paragraphs 12.1 and 13.2 for additional information.

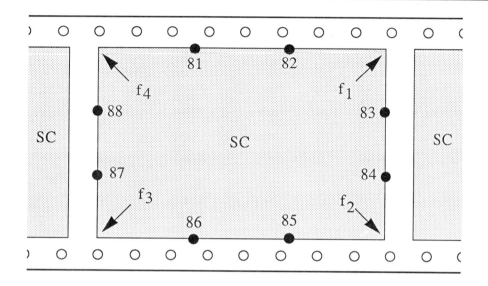

Figure E-7. Center-of-Format using Fiducials

<u>Required Values</u>

Center-of-Format			Intersecting Lines		
ID	**x**	**y**	**ID**	**x**	**y**
f_1	51.0,	36.0	A	40.000	9.000
f_2	51.0,	6.0	C	38.500	13.750
f_3	6.0,	6.0	VPX	74.300	21.000
f_4	6.0,	36.0	VPY	-5.200	21.000
			[VPDH	36.210	21.000]

Center of Format   (units = mm)

1. dx_{13} = $xf_1 - xf_3$ = 51.0 - 6.0 = 45.0
2. dy_{13} = $yf_1 - yf_3$ = 36.0 - 6.0 = 30.0
3. dx_{24} = $xf_2 - xf_4$ = 51.0 - 6.0 = 45.0
4. dy_{24} = $yf_2 - yf_4$ = 6.0 - 36.0 = -30.0
5. w_1 = $(xf_1 * yf_3) - (yf_1 * xf_3)$ = 90.0
6. w_2 = $(xf_2 * yf_4) - (yf_2 * xf_4)$ = 1,800.0
7. w_3 = $(w_1 * dx_{24}) - (w_2 * dx_{13})$ = -76,950.0
8. w_4 = $(w_1 * dy_{24}) - (w_2 * dy_{13})$ = -56,700.0
9. w_5 = $(dx_{13} * dy_{24}) - (dx_{24} * dy_{13})$ = -2,700.0
10. xo = w_3 / w_5 = 28.5
11. yo = w_4 / w_5 = 21.0

Example E-8: Center-of-Format - Two Points Per Side

The purpose is to use two points per format edge to locate the center-of-format. The E-7 procedure is used. Refer to Figure E-7 for point location, and to paragraph 12.2 for additional information.

Required Values

ID		x	y		ID		x	y
81	=	12.0,	36.0		85	=	46.0,	6.0
82	=	44.0,	36.0		86	=	12.0,	6.0
83	=	51.0,	28.0		87	=	6.0,	9.0
84	=	51.0,	12.0		88	=	6.0,	33.0

Intersection lines are A (1st) and B (2nd), and the points on each line are numbered 1 and 2.

	Compute the Corner Values			**C-1**	**C-2**	**C-3**	**C-4**
	Corner equations				_Corners_		
1.	$dx_1 = xA_1 - xA_2$	=		-32.0	0.0	34.0	0.0
2.	$dy_1 = yA_1 - yA_2$	=		0.0	16.0	0.0	-24.0
3.	$dx_2 = xB_2 - xB_1$	=		0.0	-34.0	0.0	32.0
4.	$dy_2 = yB_2 - yB_1$	=		-16.0	0.0	24.0	0.0
5.	$w_1 = (xA_1*yA_2)-(yA_1*xA_2)$	=		-1,152.0	-816.0	204.0	144.0
6.	$w_2 = (xB_2*yB_1)-(yB_2*xB_1)$	=		816.0	-204.0	-144.0	1,152.0
7.	$w_3 = (w_1*dx_2)-(w_2*dx_1)$	=		26,112.0	27,744.0	4,896.0	4,608.0
8.	$w_4 = (w_1*dy_2)-(w_2*dy_1)$	=		18,432.0	3,264.0	4,896.0	27,648.0
9.	$w_5 = (dx_1*dy_2)-(dx_2*dy_1)$	=		512.0	544.0	816.0	768.0
10.	$xc_1 = w_3 / w_5$	=		51.0	51.0	6.0	6.0
11.	$yc_1 = w_4 / w_5$	=		36.0	6.0	6.0	36.0

Example E-9: Find pp using DKA in Horizontal Plane.
The purpose of this exercise is to locate the **pp** in a cropped two-point photo, and is based on the description given in paragraph 12.3 involving a known horizontal angle. In two-point perspective the object is to locate the **CS**, and thus the **pp** and **f'**. The intersecting lines procedures of E-7/E-13 are is used to locate **VPDH**. For additional information refer to paragraph 12.3.2 and Figure 12-1.

Principal Point using DKA-Horizontal Plane Method

Required Values

ID	x	y
1.	VPX	74.300	21.000
2.	VPY	-5.200	21.000
3.	VPDH	36.210	21.000

Known angle = **ka** = 45.000°

1.	w_1	=	$x_1 - x_2$	=	79.500
2.	w_2	=	$x_3 - x_2$	=	41.410
3.	w_3	=	$(\tan(\mathbf{ka}))^2$	=	1.000
4.	w_4	=	$w_1 {}^* w_2{}^2$	=	136,325.654
5.	w_5	=	$w_3 {}^* (w_1 - w_2)^2 + w_2{}^2$	=	3,165.63616
6.	x_4	=	$(w_4 / w_5) + x_2$	=	37.8642206
7.	y_4	=	$y_2 - ((x_4 - x_2){}^* (x_1 - x_4))^{1/2}$	=	-18.6115948

Note: x_4 and y_4 are the graphical camera-station coordinates used in two-point perspective. By finding these values you have determined the location of the **pp**, and the value of **f'**.

Example E-10: Find pp using DKA in Vertical Plane - VPY.

The purpose of this exercise is to locate the **pp** in a cropped two-point photo, and is based on the description given in paragraph 12.3.3 involving a known vertical angle. Figure 12-2 contains additional information.

Vanishing Point Diagonal - Vertical at VPY Two-Point Solution

Required Values

	ID	x	y
1.	VPX	74.300	21.000
2.	VPY	-5.200	21.000
3.	A1	-5.200	21.000
4.	A2	-5.200	30.000
5.	E1	40.000	9.000
6.	E2	29.000	26.600

Known angle = **ka** = 45.000°

Line A is the perpendicular to the Horizon line. Point A_1 is **VPY**. Point A_2 x value is same as **VPYx**, and the point A_2y value is some arbitrary value above the **VPY**. Line E is the diagonal line with point E_2 closest to VPDV. The following quantities are computed in the order given.

1.	dy_1	=	$yA_1 - yA_2$		=	-9.00
2.	dx_2	=	$xB_2 - xB_1$		=	-11.00
3.	dy_2	=	$yB_2 - yB_1$		=	17.60
4.	w_1	=	$(xA_1 * yA_2) - (yA_1 * xA_2)$		=	-46.80
5.	w_2	=	$(xB_2 * yB_1) - (yB_2 * xB_1)$		=	-803.00
6.	w_3	=	$(w_1 * dx_2)$		=	514.80
7.	w_4	=	$(w_1 * dy_2) - (w_2 * dy_1)$		=	-8,050.68
8.	w_5	=	$-(dx_2 * dy_1)$		=	-99.00
9.	VPDx	=	w_3 / w_5		=	-5.20
10.	VPDy	=	w_4 / w_5		=	81.32
11.	w_6	=	Tan(**ka**)		=	1.00
12.	MPYx	=	VPYx + ((VPDy-VPYy) * w_6)		=	55.12

Note: Points **VPY** (P2), semicircle center (P0), and camera station (P4) form an isosceles triangle, where the base is the calculated distance between **VPY** and **MPY** (see Figures 12-2). The sides of the isosceles triangle are the radius of the arc through the camera station. The angle at **VPY** can now be determined.

13.	$\cos(P2)$	=	(MPYx-VPYx)/|VPXx-VPYx|	=	0.758742138
	$\sin(P2)$	=	$(1.-(\cos(P2)^2)^{1/2}$	=	0.651391102
14.	CSx	=	VPYx + (MPYx-VPYx)*cos(P2)	=	-5.2
15.	CSy	=	VPYy - (MPYx-VPYx)*sin(P2)	=	21.0
16.	xo	=	CSx	=	-5.200
17.	yo	=	VPYy (or y value of **THL**)	=	21.000

Example E-11: Find pp using DKA in Vertical Plane - VPX.
    The purpose of this exercise is the same as E-10. The exercise is left for the reader to complete.

Vanishing Point Diagonal - Vertical at VPX Two-Point Solution

<u>Required Values</u>

	ID	x	y
1.	VPX	74.300	21.000
2.	VPY	-5.200	21.000
3.	A1	-5.200	21.000
4.	A2	-5.200	30.000
5.	G1	40.000	9.000
6.	G2	29.000	26.600

Known angle = **ka** = 45.000°

1.	dy_1	=	$yA_1 - yA_2$	=	....
2.	dx_2	=	$xG_2 - xG_1$	=	....
3.	dy_2	=	$yG_2 - yG_1$	=	....
4.	w_1	=	$(xA_1 * yA_2) - (yA_1 * xA_2)$	=	....
5.	w_2	=	$(xG_2 * yG_1) - (yG_2 * xG_1)$	=	....
6.	w_3	=	$(w_1 * dx_2)$	=	....
7.	w_4	=	$(w_1 * dy_2) - (w_2 * dy_1)$	=	....
8.	w_5	=	$-(dx_2 * dy_1)$	=	....
9.	VPDx	=	$w_3 / w_5 = $ VPXx	=	....
10.	VPDy	=	w_4 / w_5	=	....
11.	w_6	=	Tan(**ka**)	=	....
12.	MPYx	=	VPYx + (VPDy-VPXy)*w_6	=	....

    For this geometry the notation changes, and becomes: points **VPX** (P1), semicircle center (P0), and camera station (P4) form an isosceles triangle, where the base is the calculated distance between **VPX** and **MPX**. The sides of the isosceles triangle are the radius of the arc through the camera station. The angle at **VPX** can now be determined.

13. $\text{Cos}(P1) =$ $(VPXx\text{-}MPXx)/|VPXx\text{-}VPYx|$ = ...

      $\text{Sin}(P1) =$ $(1.\text{-}(\text{Cos}(P1)^2)^{1/2}$ = ...

14. CSx = $VPXx\text{-}(VPXx\text{-}MPXx)*\text{Cos}(P1)$ = ...

15. CSy = $VPYy\text{-}(VPXx\text{-}MPXx)*\text{Sin}(P1)$ = ...

16. xo = CSx = _____

17. yo = VPYy (or y value of **THL**) = _____

Example E-12: Find pp or Major VPs - Orthocenter Method.

The purpose of this exercise is to locate **pp** as one of the major three-point perspective geometry points: **VPX, VPY, VPZ**, or **pp**. The photo can be full-format or cropped. Paragraph 12.4 contains more information on how any three of the major points can be used to find the fourth.

Principal Point using Orthocenter Method

<u>Required Values</u>

	ID	x	y
1.	VPX	57.1068301	36.246098
2.	VPY	-12.9471360	32.899702
3.	VPZ	29.0764351	-16.827256

1.	dx_1x_3	=	$x_1 - x_3$	=	28.030395
2.	dx_2x_3	=	$x_2 - x_3$	=	-42.023571
3.	dy_1y_3	=	$y_1 - y_3$	=	53.073354
4.	dy_2y_3	=	$y_2 - y_3$	=	49.726958
5.	w_1	=	$(y_1 * dy_2y_3) + (x_1 * dx_2x_3)$	=	-597.424732
6.	w_2	=	$(y_2 * dy_1y_3) + (x_2 * dx_1x_3)$	=	1,383.18419
7.	w_3	=	$(w_1 * dy_1y_3) - (w_2 * dy_2y_3)$	=	-100,488.877
8.	w_4	=	$(w_1 * dx_1x_3) - (w_2 * dx_2x_3)$	=	41,380.288
9.	w_5	=	$(dy_1y_3 * dx_2x_3) - (dy_2y_3 * dx_1x_3)$	=	-3,624.19814
10.	xo	=	w_3 / w_5	=	<u>27.7272028</u>
11.	yo	=	$-w_4 / w_5$	=	<u>11.4177775</u>

Example E-14: Find VP - Segmented Line Method.
    The purpose of this exercise is to locate a vanishing point when two or more parallel lines are not available. You are given a full-format photograph of an object; e.g., building face, bridge, structural tower, and so on, which has the following geometry: a line in the direction of one object-space axis, and another line (perpendicular and segmented) in the direction of another object-space axis. In this example the two object-space axes are X and Y. Points on the X-axis are labeled A, and points on the Y-axis are labeled B. More information can be found in paragraph 13.3.

Required Values

ID	x	y
1.	-84.404	31.556
2.	-102.590	31.176
3.	-120.572	32.322

$D_1 = 16.0$ $D_2 = 64.0$

Ratio K = 16.0/64.0 = 0.25

Segmented Line

1.	dx_1 =	$x_2 - x_1$	=	-18.186
2.	dx_2 =	$x_3 - x_2$	=	-17.982
3.	dy_1 =	$y_2 - y_1$	=	-0.38000
4.	dy_2 =	$y_3 - y_2$	=	1.1460
5.	K =	D_1 / D_2	=	0.2500

Note: If the ratio of K is known then D_1 and D_2 do not have to be known. D_1 and D_2 are object-space distances.

6.	w_1 =	dx_2 / dx_1	=	0.988782582
7.	w_2 =	dy_2 / dy_1	=	-3.015789390
8.	w_3 =	$(w_1 + w_2)/2.$	=	....
9.	If $|w_1| < |w_2|$, $w_3 = w_2$		=	-3.015789390
10.	If $|w_2| < |w_1|$, $w_3 = w_1$		=	....

Note: If the line is straight, and not parallel to one of the object-space axes, then w_1 should equal w_2. Errors in measuring will necessitate the average value (w_3) to be determined. When the line is parallel to one of the object-space axes, use the appropriate w_1 or w_2 values; e.g., if line is parallel to the X-axis, use $w_3 = w_2$.

11. w_4 = $K * w_3$ = -.753947347

12. w_5 = $x_1 * w_4 - x_1$ = ~~184.208172~~ 148.040179

13. w_6 = $y_1 * w_4 - y_1$ = ~~-56.1135625~~ -55.347562

14. VPX = $w_5 / (w_4 - 1.)$ = -105.024915   -84.404   } = pt. 1

15. VPY = $w_6 / (w_4 - 1.)$ = 31.9927292   31.556   } ?

Example E-15: Find VP - Orthogonal Line Method VP - (two procedures)
    In this exercise the purpose is to locate a **VPY** using a known (**VPX**), an arbitrary line B through **VPX**, and a line A perpendicular to the line B, and through the **VPY**. In this example point A_2 is intersection point of lines A and B. Additional information can be found in paragraph 13.4.

Orthogonal Line - Procedure No. 1 (Units=mm)

Required Values

ID	x	y
VPX	57.1068301	36.2460980
pp	27.7272028	11.4177775
A_1	26.2384727	27.5167257
A_2	34.8573293	26.3327426
f' =	25.7223017	

1.	w_1	=	VPXx - xo	=	29.3796273
2.	w_2	=	VPXy - yo	=	24.8283205
3.	w_3	=	$(w_1 * xo) + (w_2 * yo) - f'^2$	=	436.462318
4.	w_4	=	$yA_2 - yA_1$	=	-1.18398311
5.	w_5	=	$xA_2 - xA_1$	=	8.6188564
6.	w_6	=	$(xA_1 * yA_2) - (xA_2 * yA_1)$	=	-268.228616
7.	w_7	=	$(w_3 * w_5) + (w_6 * w_2)$	=	-2,897.860
8.	w_8	=	$(w_3 * w_4) - (w_6 * w_1)$	=	7,363.69275
9.	w_9	=	$(w_1 * w_5) + (w_4 * w_2)$	=	223.822476
10.	VPYx	=	w_7 / w_9	=	<u>-12.9471358</u>
11.	VPYy	=	w_8 / w_9	=	<u>32.8997019</u>

    In Procedure No. 2 the origin of the object-space axes (xa, ya) and one point on each of the three object-space axes (xw,yw; xu,yu; xv,yv) to the vanishing points are imaged. Point **w** is on the imaged X-axis, point **u** is on the imaged Y-axis, and point **v** is on the imaged Z-axis. The origin of the image-coordinate system is at the principal point. The following values should be computed in the order given.

Orthogonal Line - Procedure No. 2 (Units = mm)

Required Values

ID	x	y
A	32.0000000	5.0000000
W	38.8733088	13.5540101
U	27.5007818	7.79276571
V	34.8573293	26.3327426
pp	27.7272028	11.4177775
f' =	25.7223017	

Note: All x,y coordinates are given as measured and are assumed to have been translated using **pp** when used in the equations. The computed rotation matrix is not the **a,t,s** matrix, and should not be used elsewhere.

1.	da	$=$	$(xa^2 + ya^2 + f'^2)^{1/2}$	$=$	26.8529602
2.	r_{13}	$=$	$-xa / da$	$=$	-0.159118294
3.	r_{23}	$=$	$-ya / da$	$=$	0.238997022
4.	r_{33}	$=$	f' / da	$=$	0.957894457
5.	w_1	$=$	$(r_{13}*xv)+(r_{23}*yv)-(r_{33}*f')$	$=$	-22.2091516
6.	w_2	$=$	$(xv^2+ yv^2 + f'^2 - w_1^2)^{1/2}$	$=$	21.0163099
			(w_2 takes on the sign of yv - ya)		
7.	w_3	$=$	$xv - w_1 * r_{13}$	$=$	3.5962442
8.	w_4	$=$	$yv - w_1 * r_{23}$	$=$	20.2228862
9.	w_5	$=$	$-f' - w_1 * r_{33}$	$=$	-4.44827853
10.	r_{12}	$=$	w_3 / w_2	$=$	0.171116824
11.	r_{22}	$=$	w_4 / w_2	$=$	0.96224724
12.	r_{32}	$=$	w_5 / w_2	$=$	-.211658401
13.	r_{11}	$=$	$(1. - r_{12}^2 - r_{13}^2)^{1/2}$	$=$	.972317027
14.	w_6	$=$	$(r_{12}*r_{22}) + (r_{13}*r_{23})$	$=$	.126627893
15.	w_7	$=$	$(r_{12}*r_{32}) + (r_{13}*r_{33})$	$=$	-.188636845
16.	r_{21}	$=$	$-w_6 / r_{11}$	$=$	-.130233134
17.	r_{31}	$=$	$-w_7 / r_{11}$	$=$	.194007551
18.	w_8	$=$	$r_{11}*xw + r_{21}*yw - r_{31}*f'$	$=$	5.56901965
19.	w_9	$=$	$r_{12}*xw + r_{22}*yw - r_{32}*f'$	$=$	9.40721142
20.	w_{10}	$=$	$r_{11}*xu + r_{21}*yu - r_{31}*f'$	$=$	-4.7383771

21. w_{11} = $r_{12}*xu + r_{22}*yu - r_{32}*f'$ =   1.9174392

22. w_{12} = w_8 / w_9 =   .591994736

23. w_{13} = w_{10} / w_{11} =   -2.4712007

24. w_{14} = $(|w_{12} * w_{13} + 1.|)^{1/2}$ =   .680395331

     if camera points down, then $w_{14}= -w_{14}$ =   -.680395331

25. w_{15} = $r_{11}*w_{12} + r_{12} + r_{13}*w_{14}$ =   .85498673

26. w_{16} = $r_{21}*w_{12} + r_{22} + r_{23}*w_{14}$ =   .722537453

27. w_{17} = $r_{31}*w_{12} + r_{32} + r_{33}*w_{14}$ =   -.748553868

28. w_{18} = $r_{11}*w_{13} + r_{12} + r_{13}*w_{14}$ =   -2.12341035

29. w_{19} = $r_{21}*w_{13} + r_{22} + r_{23}*w_{14}$ =   1.12146699

30. w_{20} = $r_{31}*w_{13} + r_{32} + r_{33}*w_{14}$ =   -1.34283691

31. w_{21} = $r_{12} * w_{14} - r_{13}$ =   0.0426912056

32. w_{22} = $r_{22} * w_{14} - r_{23}$ =   -.893705552

33. w_{23} = $r_{32} * w_{14} - r_{33}$ =   -.813883069

34. VPXx = $-f' * w_{15} / w_{17}$ =   <u>29.3796179</u>

35. VPXy = $-f' * w_{16} / w_{17}$ =   <u>24.8283085</u>

36. VPYx = $-f' * w_{18} / w_{20}$ =   <u>-40.6743374</u>

37. VPYy = $-f' * w_{19} / w_{20}$ =   <u>21.4819180</u>

38. VPZx = $-f' * w_{21} / w_{23}$ =   <u>1.34925075</u>

39. VPZy = $-f' * w_{22} / w_{23}$ =   <u>-28.2450449</u>

Example E-16: Find VP - Use Interior Parameters and [R].

The purpose of this exercise is to determine the **VP** given interior parameters. The example is our standard cube, and is useful in accurately approximating values when some parametric values are known. Additional information can be found in paragraph 13.5.

pp, f', and [R] are known - full-format image
Required Values

$xo =$ 27.7272028, $yo =$ 11.4177775

$f' =$ 25.7223017 mm

$$[R] = \begin{bmatrix} .634911393 & -.77177828 & -.0352960473 \\ .536554920 & .40761038 & .7388928160 \\ -.555874391 & -.48806974 & .6728979080 \end{bmatrix}$$

1.	w_1	$=$	r_{11} / r_{31}	$=$	-1.1421852
2.	w_2	$=$	r_{21} / r_{31}	$=$	-.9652441
3.	w_3	$=$	r_{12} / r_{32}	$=$	1.5812869
4.	w_4	$=$	r_{22} / r_{32}	$=$	-.8351470
5.	w_5	$=$	r_{13} / r_{33}	$=$	-.0524539
6.	w_6	$=$	r_{23} / r_{33}	$=$	1.0980766
7.	VPXx	$=$	$xo - f' * w_1$	$=$	57.1068301
8.	VPXy	$=$	$yo - f' * w_2$	$=$	36.246098
9.	VPYx	$=$	$xo - f' * w_3$	$=$	-12.947136
10.	VPYy	$=$	$yo - f' * w_4$	$=$	32.899702
11.	VPZx	$=$	$xo - f' * w_5$	$=$	29.0764351
12.	VPZy	$=$	$yo - f' * w_6$	$=$	-16.8272559

Example E-17: Find [R] Using a, t, and s

The purose of this exersise is to calculate the [R], and is best left up to the reader. The values of the rotation matrix are given in E-16. Paragraph 14.2 contains more information.

<u>Required values</u>

a = 48.716167°
t = 47.7088766°
s = 177.265126°

1.	SA	=	sin(a)	=	....
2.	CA	=	cos(a)	=	....
3.	ST	=	sin(t)	=	....
4.	CT	=	cos(t)	=	....
5.	SS	=	sin(s)	=	....
6.	CS	=	cos(s)	=	....
7.	w_1	=	-SA * CT	=	....
8.	w_2	=	-CA * CT	=	....
9.	r_{11}	=	(w_1 * SS) - (CA * CS)	=	....
10.	r_{12}	=	(w_2 * SS) + (SA * CS)	=	....
11.	r_{13}	=	-ST * SS	=	....
12.	r_{21}	=	(w_1 * CS) + (CA * SS)	=	....
13.	r_{22}	=	(w_2 * CS) - (SA * SS)	=	....
14.	r_{23}	=	-ST * CS	=	....
15.	r_{31}	=	-ST * SA	=	....
16.	r_{32}	=	-ST * CA	=	....
17.	r_{33}	=	CT	=	....

Example E-18: Find [R] Using pp, f', a, and Major VP - (three procedures)

The purpose is to determine [**R**]. We have worked Procedure No. 1 as an example. Procedures No. 2 and No. 3 are left to the reader. It is sufficient to state that the largest single problem with these procedures is the problem of human error. Paragraph 14.3 contains more information.

[R] from pp, f', a, and One VP
Procedure No. 1 - VP = VPX

Required Values

ID	x	y	(units = mm)
VPX	57.1068301	36.2460980	
VPY	-12.9471360	32.8997020	
VPZ	29.0764351	-16.827256	
pp:	27.7272028	11.4177775	
f' =	25.7223017		

1.	w_1	=	VPXx - xo	=	29.3796272
2.	w_2	=	VPXy - yo	=	24.8283205
3.	VRx	=	$(w_1^2 + w_2^2 + f'^2)^{1/2}$	=	46.2735865
4.	r_{11}	=	w_1 / VRx	=	0.634911393
5.	r_{21}	=	w_2 / VRx	=	0.536554920
6.	r_{31}	=	-f' / VRx	=	-.555874391
7.	r_{32}	=	r_{31} / tan(**a**)	=	-.488069735
8.	r_{33}	=	$(1.0 - r_{31}^2 - r_{32}^2)^{1/2}$	=	0.672897908

[if camera points upward, then $r_{33} = -r_{33}$]

9.	w_3	=	$(r_{11} * r_{33}) - (r_{21} * r_{31} * r_{32})$	=	0.281660265
10.	w_4	=	$r_{11}^2 + r_{21}^2$	=	0.691003660
11.	r_{22}	=	w_3 / w_4	=	0.407610381
12.	w_5	=	$(r_{11} * r_{22}) - r_{33}$	=	-.414101434
13.	r_{12}	=	w_5 / r_{21}	=	-.771778280
14.	w_6	=	$(r_{11} * r_{31}) + (r_{12} * r_{32})$	=	0.0237506364
15.	w_7	=	$(r_{21} * r_{31}) + (r_{22} * r_{32})$	=	-.497199430
16.	r_{13}	=	$-w_6$ / r_{33}	=	-.0352960473
17.	r_{23}	=	$-w_7$ / r_{33}	=	0.7388928160

[R] from pp, f', a, and one VP
Procedure No. 2 - VP = VPY : E-18 continued

1.	w_1	=	VPYx - xo		=	....
2.	w_2	=	VPYy - yo		=	....
3.	VRy	=	$(w_1^2 + w_2^2 + f'^2)^{1/2}$		=	....
4.	r_{12}	=	w_1 / VRy		=	_____
5.	r_{22}	=	w_2 / VRy		=	_____
6.	r_{32}	=	$-f' / VRy$		=	_____
7.	r_{31}	=	$r_{32} * \tan(a)$		=	_____
8.	r_{33}	=	$(1.0 - r_{31}^2 - r_{32}^2)^{1/2}$		=	_____

[if camera points upward, then $r_{33} = -r_{33}$]

9.	w_3	=	$(r_{31} * r_{12}) - (r_{22} * r_{32} * r_{33})$		=	....
10.	w_4	=	$r_{12}^2 + r_{22}^2$		=	....
11.	r_{23}	=	w_3 / w_4		=	_____
12.	w_5	=	$(r_{12} * r_{23}) - r_{31}$		=	....
13.	r_{13}	=	w_5 / r_{22}		=	_____
14.	w_6	=	$(r_{12} * r_{32}) + (r_{13} * r_{33})$		=	....
15.	w_7	=	$(r_{22} * r_{32}) + (r_{23} * r_{33})$		=	....
16.	r_{11}	=	$-w_6 / r_{31}$		=	_____
17.	r_{21}	=	$-w_7 / r_{31}$		=	_____

[R] from pp, f', a, and one VP
Procedure No. 3 - VP = VPZ : E-18 continued

1.	w_1	=	VPZx - xo		=	....
2.	w_2	=	VPZy - yo		=	....
3.	VRz	=	$(w_1^2 + w_2^2 + f'^2)^{1/2}$		=	....
4.	r_{13}	=	$-w_1 / VRz$		=	_____
5.	r_{23}	=	$-w_2 / VRz$		=	_____
6.	r_{33}	=	f' / VRz		=	_____

[if camera points up, then $r_{13} = -r_{13}$, $r_{23} = -r_{23}$, $r_{33} = -r_{33}$]

7.	w_3	=	$r_{13}^2 + r_{23}^2$		=	....
8.	w_4	=	$-(w_3)^{1/2}$		=	....
9.	r_{31}	=	$w_4 * \sin(a)$		=	_____
10.	r_{32}	=	$w_4 * \cos(a)$		=	_____

11. w_5 = $(r_{32} * r_{13}) - (r_{23} * r_{31} * r_{33})$ = ...
12. r_{21} = w_5 / w_3 = _____
13. w_6 = $(r_{21} * r_{13}) - r_{32}$ = ...
14. r_{11} = w_6 / r_{23} = _____
15. w_7 = $(r_{11} * r_{31}) + (r_{13} * r_{33})$ = ...
16. w_8 = $(r_{21} * r_{31}) + (r_{23} * r_{33})$ = ...
17. r_{12} = $-w_7 / r_{32}$ = _____
18. r_{22} = $-w_8 / r_{32}$ = _____

Example E-19: Find [R] Using pp, f', and VPX, VPY, VPZ

The purpose is to find [R], and is another one of those examples best completed by the reader. Results should equal those in E-18, and paragraph 14.4 contains more information.

pp, f', VPX, VPY, and VPZ Method

<u>Required Values</u>

ID	x	y	(units = mm)
VPX	**57.1068301**	**36.2460980**	
VPY	**-12.9471360**	**32.8997020**	
VPZ	**29.0764351**	**-16.8272559**	
pp	27.7272028	11.4177775	
f' =	**25.7223017**		

1. w_1 = VPXx - xo = ...
2. w_2 = VPYx - xo = ...
3. w_3 = VPZx - xo = ...
4. w_4 = VPXy - yo = ...
5. w_5 = VPYy - yo = ...
6. w_6 = VPZy - yo = ...
7. VRx = $(w_1^2 + w_4^2 + f'^2)^{1/2}$ = ...
8. VRy = $(w_2^2 + w_5^2 + f'^2)^{1/2}$ = ...
9. VRz = $(w_3^2 + w_6^2 + f'^2)^{1/2}$ = ...
10. r_{11} = w_1 / VRx = _____
11. r_{21} = w_4 / VRx = _____
12. r_{31} = -f' / VRx = _____
13. r_{12} = w_2 / VRy = _____
14. r_{22} = w_5 / VRy = _____
15. r_{23} = -f' / VRy = _____
16. r_{13} = w3 / VRz = _____
17. r_{23} = w_6 / VRz = _____
18. r_{33} = -f' / VRz = _____

Note: If the values of w_6 is negative then: $r_{13} = -r_{13}$, $r_{23} = -r_{23}$, and $r_{33} = -r_{33}$

Example E-20: Find f' Using pp and two VPs - Three Procedures

The purpose of this exercise is to determine the **f'** value. These are repetitive procedures, and the example is Procedure No. 1. Procedures No. 2 and No. 3 have been left for the reader to complete. Paragraph 15.3 contains more information.

Procedure No. 1: **f' Using pp, VPX, and VPY**

<u>Required Values</u>

ID	x	y	(units = mm)
VPX:	**57.1068301**	**36.2460980**	
VPY:	**-12.9471360**	**32.8997020**	
pp:	**27.7272028**	**11.4177775**	

1.	w_1	=	VPXx - xo	=	29.3796272
2.	w_2	=	VPYx - xo	=	-40.6743388
3.	w_3	=	VPXy - yo	=	24.8283205
4.	w_4	=	VPYy - yo	=	21.4819245
5.	w_5	=	$(w_1 * w_2) + (w_3 * w_4)$	=	-661.636808
6.	**f'**	=	$(|w_5|)^{1/2}$	=	<u>25.7223018</u>

Procedure No. 2: **f' Using pp, VPX, and VPZ**

<u>Required Values</u>

ID	x	y	(units = mm)
VPX	**57.1068301**	**36.2460980**	
VPZ	**29.0764351**	**-16.8272559**	
pp	**27.7272028**	**11.4177775**	

1.	w_1	=	VPXx - xo	=	....
2.	w_2	=	VPZx - xo	=	....
3.	w_3	=	VPXy - yo	=	....
4.	w_4	=	VPZy - yo	=	....
5.	w_5	=	$(w_1 * w_2) + (w_3 * w_4)$	=	....
6.	**f'**	=	$(|w_5|)^{1/2}$	=	_____

Procedure No. 3: **f' Using pp, VPY, and VPZ**

<u>Required Values</u>

ID	x	y	(units = mm)
VPY	-12.9471360	32.8997020	
VPZ	29.0764351	-16.8272559	
pp	27.7272028	11.4177775	

1. w_1 =   VPYx - xo   = ...
2. w_2 =   VPZx - xo   = ...
3. w_3 =   VPYy - yo   = ...
4. w_4 =   VPZy - yo   = ...
5. w_5 = $(w_1 * w_2) + (w_3 * w_4)$ = ...
6. **f'**   = $(|w_5|)^{1/2}$ =   _____

Example E-21: Find a, t, and s from [R]

The purpose of this exercise is to determe the rotation angles, having determined the [R]. (We are still using one of the standard rotation matrices; however, it would be best if you checked the results by using these angles to compute the [R].) Paragraph 16.2 contains more information.

Required Values

$$r_{13} = -.0352960473 \qquad r_{23} = -.738892816$$
$$r_{31} = -.555743910 \qquad r_{32} = 0.488069735$$
$$r_{33} = 0.672897908$$

Azimuth, Tilt, and Swing from [R]

1.	w_1	$=$	r_{31} / r_{32}		$=$	-1.138922412
2.	w_2	$=$	r_{13} / r_{23}		$=$	-.0477688324
3.	w_3	$=$	$(r_{13}^2 + r_{23}^2)^{1/2}$		$=$	0.7397353610
4.	a	$=$	$\text{Tan}^{-1}(|w1|)$	[0° to 90°]	$=$	<u>48.716167°</u>
4.	t	$=$	$\text{Cos}^{-1}(r_{33})$	[if r_{33} neg.]	$=$	<u>-</u>
5.	t	$=$	$\text{Sin}^{-1}(w_3)$	[if r_{33} pos.]	$=$	<u>47.7088766°</u>
			[where t = 0° to 180°]			
6.	s	$=$	$\text{Tan}^{-1}(w_2) + 180°$	[90° to 270°]	$=$	<u>177.265126°</u>

Example E-22: Find a, t, and s from pp, f', and the Three Major VPs

The purpose of this exercise is to determine the rotation angles. This is basically a three-point perspective exercise; however, it is easy to use two-point perspective parameters, since it is known that **t** is 90°, and image coordinates defining a direction for **VPZ** are definable. Paragraph 16.3 contains more information.

Using pp, f', VPX, VPY, and VPZ

<u>Required Values</u>

ID	x	y	(units = mm)
VPX:	57.1068301	36.2460980	
VPY:	-12.9471360	32.8997020	
pp:	27.7272028	11.4177775	
f' =	25.7223017		

1.	w_1	=	VPXx - xo	=	29.3796272
2.	w_2	=	VPYx - xo	=	-40.6743388
3.	w_3	=	VPZx - xo	=	1.3492322
4.	w_4	=	VPXy - yo	=	24.8283205
5.	w_5	=	VPYy - yo	=	21.4819245
6.	w_6	=	VPZy - yo	=	-28.2450335
7.	VRx	=	$(w_1^2+w_4^2+f'^2)^{1/2}$	=	46.2735865
8.	VRy	=	$(w_2^2+w_5^2+f'^2)^{1/2}$	=	52.7021036
9.	VRz	=	$(w_3^2+w_6^2+f'^2)^{1/2}$	=	38.226158
10.	w_7	=	VRy / VRx	=	1.13892412
11.	w_8	=	f' / VRz	=	0.672897907
12.	w_9	=	$(w_3^2+w_6^2)^{1/2}$ / VRz	=	0.739735362
13.	w_{10}	=	w_3 / w_6	=	-.0477688298
14.	**a**	=	Tan^{-1} ($|w_7|$)	=	<u>48.7161671</u>
15.	**t**	=	Cos^{-1} (w_8) [if w_6 is neg.]	=	<u> . </u>
16.	**t**	=	Sin^{-1} (w_9) [if w_6 is pos.]	=	<u>47.7088766</u>
17.	**s**	=	Tan^{-1} (w_{10}) + 180°	=	<u>177.265127</u>

Example E-23: Find One CS Coordinate (Xc, Yc, or Zc)

The purpose of this exercise is to determine the object-space, or model-space, camera station coordinates. The Phase One parameters must be known.

Method for Calculating One Object-Space CS Coordinate

The procedures for determining Xc, Yc, or Zc are identical until Step 11, at which time the plane of the given distance determines the completion of the solution. See Chapter Eighteen, paragraph 18.2 for additional information.

Given distance parallel to a major plane: DXY, DXZ, DYZ

The end point of the known dimension with the known object-space coordinate is point number 1 in the solution. Using whatever means available, the image points have been measured and recorded with known units. All of the Phase One parametric values for the image are known. Although it is possible to compute all three **CS** object-space coordinates with these procedures, it is not advisable to do so because of certain weaknesses in geometry that can provide incorrect results.

<u>Required Values</u>

ID	x	y (units = mm)
VPX	**57.1068301**	**36.246098**
VPY	**-12.9471360**	**32.899702**
VPZ	**29.0764351**	**-16.8272559**
pp	**27.7272028**	**11.4177775**
1	32.0000000	5.0000000
2	27.5007827	7.79276571
3	34.8573293	26.3327426

f' = 25.7223017

$$[R] = \begin{bmatrix} .634911393 & -.771778280 & -.035296047 \\ .536554920 & .407610381 & .738892816 \\ -.555874391 & -.488069735 & .672897908 \end{bmatrix}$$

1. w_1 = $x_1 - xo$ =   4.2727972
2. w_2 = $x_2 - xo$ =   -.22642009
3. w_3 = $y_1 - yo$ =   -6.4177775
4. w_4 = $y_2 - yo$ =   -3.6250118
5. w_5 = $r_{11}*w_1 + r_{21}*w_3 - r_{31}*f'$ =   13.5677263
6. w_6 = $r_{12}*w_1 + r_{22}*w_3 - r_{32}*f'$ =   6.64067216
7. w_7 = $r_{13}*w_1 + r_{23}*w_3 - r_{33}*f'$ =   -22.2013455
8. w_8 = $r_{11}*w_2 + r_{21}*w_4 - r_{31}*f'$ =   12.2095942
9. w_9 = $r_{12}*w_2 + r_{22}*w_4 - r_{32}*f'$ =   11.2514306
10. w_{10} = $r_{13}*w_2 + r_{23}*w_4 - r_{33}*f'$ =   -19.9789865

Case One: The given distance is parallel to the XY plane:
DXY = known distance = <u>24.500 ft</u>

11. w_{11} = w_5 / w_7 =   -.611121803
12. w_{12} = w_6 / w_7 =   -.299111247
13. w_{13} = w_8 / w_{10} =   -.611121801
14. w_{14} = w_9 / w_{10} =   -.563163235
15. w_{15} = $w_{13} - w_{11}$ = $2.32830644*10^{-9}$
16. w_{16} = $w_{14} - w_{12}$ =   -.264051987
17. w_{17} = $(w_{15}{}^2 + w_{16}{}^2)^{1/2}$ =   .264051988

By definition Z_1 is greater than Zc ($Z_1 > Zc$), if not then $w_{17} = -w_{17}$. $Z_1 < Zc$, therefore $w_{17} = -.264051988$

(For this example point 1 is below the **THL.**)

18.   Given X_1: Xc = $X1-(w_{11}*(DXY/w_{17}))$ =   <u>-56.7027891</u>
       Given Y_1: Yc = $Y1-(w_{12}*(DXY/w_{17}))$ =   <u>-27.7529650</u>
       Given Z_1: Zc = $Z1 - (DXY / w_{17})$ =   <u>92.7847589</u>

Case Two: The given distance is parallel to the XZ plane:
DXZ = known distance = <u>78.963 ft</u>

11. w_{11} $=$ w_5 / w_6 $=$ ....
12. w_{12} $=$ w_7 / w_6 $=$ ....
13. w_{13} $=$ w_8 / w_9 $=$ ....
14. w_{14} $=$ w_{10} / w_9 $=$ ....
15. w_{15} $=$ $w_{13} - w_{11}$ $=$ ....
16. w_{16} $=$ $w_{14} - w_{12}$ $=$ ....
17. w_{17} $=$ $(w_{15}^2 + w_{16}^2)^{1/2}$ $=$ ....

By definition, $Y_1 > Yc$: if not then $w_{17} = -w_{17}$

18.   Given X_1: $Xc = X_1 - (w_{11} * (DXZ / w_{17}))$ $=$ <u>-56.7027891</u>
      Given Y_1: $Yc = Y_1 - (DXZ / w_{17})$ $=$ <u>-27.7529650</u>
      Given Z_1: $Zc = Z_1 - (w_{12} * (DXZ / w_{17}))$ $=$ <u>92.7847589</u>

Case Three: The given distance is parallel to the YZ plane:
DYZ = known distance = <u>78.963 ft</u>

11. w_{11} $=$ w_6 / w_5 $=$ ....
12. w_{12} $=$ w_7 / w_5 $=$ ....
13. w_{13} $=$ w_9 / w_8 $=$ ....
14. w_{14} $=$ w_{10} / w_8 $=$ ....
15. w_{15} $=$ $w_{13} - w_{11}$ $=$ ....
16. w_{16} $=$ $w_{14} - w_{12}$ $=$ ....
17. w_{17} $=$ $(w_{15}^2 + w_{16}^2)^{1/2}$ $=$ ....

By definition, $X_1 > Xc$: if not then $w_{17} = -w_{17}$

18.   Given X_1: $Xc = X_1 - (DYZ / w_{17})$ $=$ <u>-56.7027891</u>
      Given Y_1: $Yc = Y_1 - (w_{11} * (DYZ / w_{17}))$ $=$ <u>-27.7529650</u>
      Given Z_1: $Zc = Z_1 - (w_{12} * (DYZ / w_{17}))$ $=$ <u>92.7847589</u>

Example E-24: Find Two CS Coordinates (XcYc, XcZc, or YcZc)

The purpose of this exercise is to determine the remaining two object-space, or model-space, camera-station coordinates. The procedures are identical until Step 6, where the given **CS** coordinate determines the solution. The end point of the known dimension, with the known object-space coordinates, is point number 1. All Phase One parametric values for the image are known. Paragraph 18.3 contains additional information.

Required Values

ID	x	y	(units = mm)
VPX	57.1068301	36.246098	
VPY	-12.9471360	32.899702	
VPZ	29.0764351	-16.8272559	
pp	27.7272028	11.4177775	
1	32.0000000	5.0000000	
2	27.5007827	7.79276571	

$$f' = 25.7223017$$

$$[R] = \begin{bmatrix} .634911393 & -.771778280 & -.035296047 \\ .536554920 & .407610381 & .738892816 \\ -.555874391 & -.488069735 & .672897908 \end{bmatrix}$$

1. $w_1 = x_1 - xo$ $\qquad$ = $\qquad$ 4.2727972
2. $w_2 = y_1 - yo$ $\qquad$ = $\qquad$ -6.4177775
3. $w_3 = r_{11}{}^*w_1 + r_{21}{}^*w_3 - r_{31}{}^*f'$ $\qquad$ = $\qquad$ 13.5677263
4. $w_4 = r_{12}{}^*w_1 + r_{22}{}^*w_3 - r_{32}{}^*f'$ $\qquad$ = $\qquad$ 6.64067216
5. $w_5 = r_{13}{}^*w_1 + r_{23}{}^*w_3 - r_{33}{}^*f'$ $\qquad$ = $\qquad$ -22.2013455

Case One where known **CS** coordinate (Xc) $\qquad$ = $\qquad$ <u>-56.7027891</u>

6. $Yc = Y_1 + (w_4 / w_3) * (Xc - X_1)$ $\qquad$ = $\qquad$ <u>-27.752965</u>
7. $Zc = Z_1 + (w_5 / w_3) * (Xc - X_1)$ $\qquad$ = $\qquad$ <u>92.7847588</u>

Case Two where known CS coordinate (Yc)   =   -27.752965

8.   Xc   = $X_1 + (w_3 / w_4) * (Yc - Y_1)$ =   -56.7027892
9.   Zc   = $Z_1 + (w_5 / w_4) * (Yc - Y_1)$ =   92.784759

Case Three where known CS coordinate (Zc)   =   92.7847589

10.   Xc   = $X_1 + (w_3 / w_5) * (Zc - Z_1)$ =   -56.7027892
11.   Yc   = $Y_1 + (w_4 / w_5) * (Zc - Z_1)$ =   -27.752965

Example E-25: Find Three CS Coordinates (Xc, Yc, and Zc)

The purpose of this exercise is to determine the three object-space, or model-space, camera-station coordiantes. The end points of the known dimension are numbered 1 and 2 (with point number 1 being the origin, if that solution is desired). The object-space coordinates of the points are known. All Phase One parametric values for the image are known. Paragraph 18.4 contains additional information.

Required Values

ID	x	y (units = mm)
VPX	57.1068301	36.246098
VPY	-12.9471360	32.899702
VPZ	29.0764351	-16.8272559
pp	27.7272028	11.4177775
1	32.0000000	5.0000000
2	27.5007827	7.79276571

f' = 25.7223017

$$[R] = \begin{bmatrix} .634911393 & -.771778280 & -.035296047 \\ .536554920 & .407610381 & .738892816 \\ -.555874391 & -.488069735 & .672897908 \end{bmatrix}$$

1.	w_1	$= x_1 - xo$	$=$	4.2727972
2.	w_2	$= x_2 - xo$	$=$	-.22642009
3.	w_3	$= y_1 - yo$	$=$	-6.4177775
4.	w_4	$= y_2 - yo$	$=$	-3.6250118
5.	w_5	$= r_{11}*w_1 + r_{21}*w_3 - r_{31}*f'$	$=$	13.5677263
6.	w_6	$= r_{12}*w_1 + r_{22}*w_3 - r_{32}*f'$	$=$	6.64067216
7.	w_7	$= r_{13}*w_1 + r_{23}*w_3 - r_{33}*f'$	$=$	-22.2013455
8.	w_8	$= r_{11}*w_2 + r_{21}*w_4 - r_{31}*f'$	$=$	12.2095942
9.	w_9	$= r_{12}*w_2 + r_{22}*w_4 - r_{32}*f'$	$=$	11.2514306
10.	w_{10}	$= r_{13}*w_2 + r_{23}*w_4 - r_{33}*f'$	$=$	-19.9789865
11.	w_{11}	$= -w_5 * w_9$	$=$	152.656332
12.	w_{12}	$= w_8 * w_6$	$=$	81.0799122
13.	w_{13}	$= w_6 * w_9$	$=$	74.7170622

14.	w_{14}	$=$	$X_2 - X_1$		$=$	0.0
15.	w_{15}	$=$	$(w_{12}*Y_2)-(w_{11}*Y_1)-(w_{13}*w_{14})$		$=$	1,986.45785
16.	w_{16}	$=$	$w_{12} - w_{11}$		$=$	-71.5764195
17.	Yc	$=$	w_{15} / w_{16}		$=$	<u>-27.7529648</u>
18.	w_{17}	$=$	$(Y_1-Yc) * (w_5/w_6)$		$=$	56.7027888
19.	Xc	$=$	$X_1 - w_{17}$		$=$	<u>-56.7027888</u>
20.	w_{18}	$=$	$(X_1 - Xc) * (w_7/w_5)$		$=$	-92.7847583
21.	Zc	$=$	$Z_1 - w_{18}$		$=$	<u>92.7847583</u>

Example E-26: Object-Space Dimensions

The purpose of this exercise is to determine a dimension in object- or model-space. The Phase One parametric values are required, as are the image measurements of the required points. The **pp** does not have to be the image coordinate origin. Equations 1 through 11 are the same no matter which **CS** coordinate is known. Complete the calculations for the initial parametric values in the order given. Paragraph 19.1 contains additional information.

Required Values

ID	x	y	(units = mm)
pp	**27.7272028**	**11.4177775**	
1	32.0000000	5.0000000	
2	27.5007827	7.79276571	

ID	X	Y	Z	(units = feet)
CS	**-56.7027888**	**-27.7529648**	**92.7847503**	

$$f' = 25.7223017$$

$$[R] = \begin{bmatrix} .634911393 & -.771778280 & -.035296047 \\ .536554920 & .407610381 & .738892816 \\ -.555874391 & -.488069735 & .672897908 \end{bmatrix}$$

Initial Parametric Values

1.	w_1	=	$x_1 - xo$	=	4.27280
2.	w_2	=	$x_2 - xo$	=	-.226419
3.	w_3	=	$y_1 - yo$	=	-6.41777
4.	w_4	=	$y_2 - yo$	=	-3.62500
5.	w_5	=	$r_{11}*w_1 + r_{21}*w_3 - r_{31}*f'$	=	13.56780
6.	w_6	=	$r_{12}*w_1 + r_{22}*w_3 - r_{32}*f'$	=	6.64065
7.	w_7	=	$r_{13}*w_1 + r_{23}*w_3 - r_{33}*f'$	=	-22.2013
8.	w_8	=	$r_{11}*w_2 + r_{21}*w_4 - r_{31}*f'$	=	12.2096
9.	w_9	=	$r_{12}*w_2 + r_{22}*w_4 - r_{32}*f'$	=	11.2514
10.	w_{10}	=	$r_{13}*w_2 + r_{23}*w_4 - r_{33}$	=	-19.9790

Once the initial parametric values have been calculated the following calculations are required for the dimension according to the major plane paralleled.

Case One: Given a CS Coordinate Calculate a Distance Parallel to the XY Plane

11.	w_{11}	=	w_5 / w_7	=	-0.61124
12.	w_{12}	=	w_6 / w_7	=	-0.299111
13.	w_{13}	=	w_8 / w_{10}	=	-0.61124
14.	w_{14}	=	w_9 / w_{10}	=	-.563163
15.	w_{15}	=	$((w_{13}-w_{11})^2+(w_{14}-w_{12})^2)^{1/2}$	=	0.264052

Calculate the Distance Using Xc

16a.   Dist.   = $|(X_1 - Xc) * w_{15} / w_{11}|$ =   <u>24.49999</u>

Calculate the Distance Using Yc

16b.   Dist.   = $|(Y_1 - Yc) * w_{15} / w_{12}|$ =   <u>24.50000</u>

Calculate the Distance Using Zc

16c.   Dist.   = $|(Z_1 - Zc) * w_{15}|$ =   <u>24.50000</u>

To check the results compute the distance using at least two of the equations above (16a through 16c).

Case Two: Given a CS Coordinate Calculate a Distance Parallel to the XZ Plane (exercise for reader)

11.	w_{11}	=	w_5 / w_6	=	....
12.	w_{12}	=	w_7 / w_6	=	....
13.	w_{13}	=	w_8 / w_9	=	....
14.	w_{14}	=	w_{10} / w_9	=	....
15.	w_{15}	=	$((w_{13}-w_{11})^2+(w_{14}-w_{12})^2)^{1/2}$	=	....

Calculate the Distance Using Xc

16a.   Dist.   = $|(X_1 - Xc) * w_{15}|$ = ...

Calculate the Distance Using Yc

16b. Dist.   = $|(Y_1 - Yc) * w_{15} / w_{11}|$ = ...

Calculate the Distance Using Zc

16c. Dist.   = $|(Z_1 - Zc) * w_{15} / w_{12}|$ = ...

To check the results compute the distance using at least two of the equations above (16a through 16c).

Case Three: Given a CS Coordinate Calculate a Distance Parallel to the YZ Plane (exercise for reader)

11. w_{11} = w_6 / w_5 = ...
12. w_{12} = w_7 / w_5 = ...
13. w_{13} = w_9 / w_8 = ...
14. w_{14} = w_{10} / w_8 = ...
15. w_{15} = $((w_{13}-w_{11})^2+(w_{14}-w_{12})^2)^{1/2}$ = ...

Calculate the Distance Using Xc

16a. Dist.   = $|(X_1 - Xc) * w_{15} / w11|$ = ...

Calculate the Distance Using Yc

16b. Dist.   = $|(Y_1 - Yc) * w_{15}|$ = ...

Calculate the Distance Using Zc

16c. Dist.   = $|(Z_1 - Zc) * w_{15} / w_{12}|$ = ...

To check the results compute the distance using at least two of the equations above (16a through 16c).

Example E-27: Object-Space Coordinates

The purpose of this exercise is to determine two object- or model-space coordinates with the third coordinate known. Given X, Y, or Z the computations for the other two coordinates (YZ, XZ, or XY) are the same until Step 6. The ID number for the point in question is 2 for these equations. The image-space coordinates of the point are known. Compute the values in the order given. A partial list of the object-space coordinates for all nine points is given. To become familiar with the procedure compute the coordinates not given (-). Paragraph 20.2 contains additional information.

Required Values

ID	x	y	(units = mm)
pp	27.7272028	11.4177775	
1	32.0000000	5.0000000	
2	27.5007827	7.7927657	
3	35.0000000	15.0000000	
4	38.8733088	13.5540101	
5	26.2384727	27.5167257	
6	37.9457466	30.8274672	
7	44.3591740	30.5663281	
8	34.8573299	26.3327426	
9	30.0000000	27.0000000	

ID	X	Y	Z
CS	-56.7027888	-27.7529648	92.7847503
1	0.0	0.0	0.0
2	0.0	24.500	0.0
3	(-)	24.500	0.0
4	(-)	0.0	0.0
5	0.0	24.500	(-)
6	(-)	24.500	(-)
7	(-)	0.0	(-)
8	0.0	0.0	(-)
9	0.0	(-)	(-)

$$f' \quad = \quad 25.7223017$$

$$[\mathbf{R}] \quad = \quad \begin{bmatrix} .634911393 & -.771778280 & -.035296047 \\ .536554920 & .407610381 & .738892816 \\ -.555874391 & -.488069735 & .672897908 \end{bmatrix}$$

The Initial Computations for Point 1

1.	w_1	$=$	$x_2 - xo$	$=$	4.27280
2.	w_2	$=$	$y_2 - yo$	$=$	-6.41777
3.	w_3	$=$	$r_{11}*w_1 + r_{21}*w_2 - r_{31}*f'$	$=$	13.5678
4.	w_4	$=$	$r_{12}*w_1 + r_{22}*w_2 - r_{32}*f'$	$=$	6.64065
5.	w_5	$=$	$r_{13}*w_1 + r_{23}*w_2 - r_{33}*f'$	$=$	-22.2013

Case One - the Known Coordinate is Z_2

6.	X_2	$=$	$Xc + ((w_3 / w_5) * (Z_1 - Zc))$	$=$	0.00000
7.	Y_2	$=$	$Yc + ((w_4 / w_5) * (Z_1 - Zc))$	$=$	24.50000

Case Two - the Known Coordinate is Y_2

6.	X_2	$=$	$Xc + ((w_3 / w_4) * (Y_1 - Yc))$	$=$	0.00000
7.	Z_2	$=$	$Zc + ((w_5 / w_4) * (Y_1 - Yc))$	$=$	0.00000

Case Three - the Known Coordinate is X_2

6.	Y_2	$=$	$Yc + ((w_4 / w_3) * (X_1 - Xc))$	$=$	24.50000
7.	Z_2	$=$	$Zc + ((w_5 / w_3) * (X_1 - Xc))$	$=$	0.00000

Example E-28: Determining the Two-Point THL
The purpose of this exercise is to locate the **THL** so that **VPY** can be located in a cropped two-point perspective image. The method is to determine the coordinates of the intersection of a vertical line and the perpendicular to the vertical line through **VPX**. The points on the vertical line are numbered 1 and 2, **VPX** is point number 3, and the intersection is point number 4. Paragraph 21.3 contains additional information.

<u>Required Values</u>

ID	x	y
1	20.000	10.000
2	20.000	13.000
3	74.300	21.000

1.	dx	=	$x_2 - x_1$	=	0.000
2.	dy	=	$y_2 - y_1$	=	3.000
3.	w_1	=	$dx_2 + dy_2$ $dx^2 + dy^2$	=	9.000
4.	w_2	=	$y_2 * x_1$	=	~~26.000~~ 260.000
5.	w_3	=	$x_2 * y_1$	=	200.000
6.	w_4	=	$(w_3 - w_2) / w_1$	=	-6.6666667
7.	w_5	=	$(x_3 * dx + y_3 * dy) / w_1$	=	7.000
8.	x_4	=	$w_5 * dx - w_4 * dy$	=	<u>20.000</u>
9.	y_4	=	$w_5 * dy + w_4 * dx$	=	<u>21.000</u>

The line defined by this new point (point 4) and the **VPX** is the **THL**. It is now possible to locate **VPY** graphically, or analytically, by extending a top line of the crate to the **THL**.

Example E-29: Locate the CS Using Multiple Two-Point Perspective.

The purpose of this exercise is to determine Phase One camera station using two non-parallel objects resting on the same horizontal plane. This is one of the easiest two-point solutions. For example, two large crates on the loading dock rest on the same horizontal plane, but the corresponding sides of each crate are not parallel. Paragraph 21.4 contains additional information.

Required Values	(given)		(translated)	
ID	**x**	**y**	**x**	**y**
1. VPX1	74.300	21.000	79.500	0.0
2. VPY1	-5.200	21.000	0.0	0.0
3. VPX2	154.300	21.000	159.500	0.0
4. VPY2	1.500	21.000	6.700	0.0

Locate the CS using Multiple Two-Point Perspective

1.	w_1	=	VPX_1x	=	79.500
2.	w_2	=	VPX_1y	=	0.000
3.	w_3	=	$VPX_2x + VPY_2x$	=	166.200
4.	w_4	=	$VPX_2y + VPY_2y$	=	0.000
5.	C_1x	=	$w_1 / 2.$	=	39.750
6.	C_1y	=	$w_2 / 2.$	=	0.000
7.	C_2x	=	$w_3 / 2.$	=	83.1
8.	C_2y	=	$w_4 / 2.$	=	0.000
9.	w_5	=	$w_1{}^2 + w_2{}^2$	=	6,320.250
10.	R_1	=	$w_5{}^{1/2} / 2.$	=	39.750
11.	w_6	=	$(VPX_2x - VPY_2x)^2$	=	23,347.840
12.	w_7	=	$(VPX_2y - VPY_2y)^2$	=	0.000
13.	R_2	=	$(w_6 + w_7)^{1/2} / 2.0$	=	76.400
14.	w_8	=	$(C_1x - C_2x)$	=	-43.350
15.	w_9	=	$(C_1y - C_2y)$	=	0.000
16.	D_{12}	=	$(w_8{}^2 + w_9{}^2)^{1/2}$	=	43.350
17.	w_{10}	=	$Tan^{-1}(-w_9 / w_8) * R$ [Ang1]	=	0.000

(where R = 57.29578 degrees/radian)

18.	w_{11}	=	$D_{12}{}^2 + R_1{}^2 - R_2{}^2$	=	-2,377.675
19.	w_{12}	=	$R_1 * D_{12}$	=	1,723.1625

20. w_{13} = $w_{11} / (2. * w_{12})$ =   -.68991607
21. w_{14} = $Cos^{-1} (w_{13}) * R$ [Ang2]   =   133.6234655
22. w_{15} = $w_{10} + w_{14}$ [Ang3]   =   133.6234655
23. CSx = $C_1x + R_1 * Cos(W_{15}/R) = CSXYx$ =   <u>12.325836</u>
24. CSy = $C_1y - R_1 * Sin(W_{15}/R) = CSXYy$ =   <u>-28.774602</u>

Translate back to original coordinate system.
25. CSx = 12.325836 -5.2000   =   <u>7.125836</u>
26. CSy = -28.774602 + 21.00   =   <u>-7.774602</u>

Example E-30: Transformation from Three-Point into Two-Point Object-Space Coordinates.

The purpose of this exercise is to complete a transformation between three- and two-point object- or model-space coordinates. This procedure is helpful when the image contains combined two- and three-point perspective, and the two-point is the dominant geometry. To complete the transformation complete the equations as shown, and refer to Chapter Twenty-one, paragraph 21.5, for additional information.

Transformation of Object-Space Coordinates

Three-Point to Two-Point
Required Values

ID	X	Y	Z	(units = feet)
CS_2	-136.1328598	-125.4042161	64.6256421	
CS_3	-121.1629215	-59.6775785	72.0573561	

$$[R] = \begin{bmatrix} .919496861 & -.393097345 & 0.000000000 \\ 0.000000000 & 0.000000000 & 1.000000000 \\ -.393097345 & -.9194968661 & 0.000000000 \end{bmatrix}$$

$$[R]_3 = \begin{bmatrix} .953900435 & -.279795393 & -.108574892 \\ .022205623 & .426573254 & -.904180375 \\ -.299300635 & -.860087087 & -.413121468 \end{bmatrix}$$

(As stated $[M]=[R]_{2p}$, $[M]^T=[R]^T_{2p}$, and $[C]=[M]^T[R]_{3p}$)

1.	c_{11}	=	$m_{11}r_{11} + m_{21}r_{21} + m_{31}r_{31}$	=	.994762741
2.	c_{12}	=	$m_{11}r_{12} + m_{21}r_{22} + m_{31}r_{32}$	=	.358515951
3.	c_{13}	=	$m_{11}r_{13} + m_{21}r_{23} + m_{31}r_{33}$	=	-.437602944
4.	c_{21}	=	$m_{12}r_{11} + m_{22}r_{21} + m_{32}r_{31}$	=	-.099769734
5.	c_{22}	=	$m_{12}r_{12} + m_{22}r_{22} + m_{32}r_{32}$	=	.782118405
6.	c_{23}	=	$m_{12}r_{13} + m_{22}r_{23} + m_{32}r_{33}$	=	-.262209602
7.	c_{31}	=	$m_{13}r_{11} + m_{23}r_{21} + m_{33}r_{31}$	=	.022205623
8.	c_{32}	=	$m_{13}r_{12} + m_{23}r_{22} + m_{33}r_{32}$	=	.426573253
9.	c_{33}	=	$m_{13}r_{13} + m_{23}r_{23} + m_{33}r_{33}$	=	.904180375

Steps 10 through 12 are the translation of the three-point object-space coordinates from the three-point **CS**. Since we are converting the three-point object-space coordinate system into the two-point object-space coordinate system, the example is CS_3 to CS_2 object-space coordinates.

10. $\quad w_1 \quad = \quad CSX_3 - CSX_3 \qquad\qquad = \qquad 0.000$
11. $\quad w_2 \quad = \quad CSY_3 - CSY_3 \qquad\qquad = \qquad 0.000$
12. $\quad w_3 \quad = \quad CSZ_3 - CSZ_3 \qquad\qquad = \qquad 0.000$

Now complete the transformation from the three-point object-space coordinate system to the two-point object-space coordinate system. The results of Steps 13 through 15 are two-point object-space coordinates for the **CS**.

13. $\quad CSX_3 \quad = \quad Xc_2p(w_1c_{11} + w_2c_{12} + w_3c_{13}) \qquad = \qquad \underline{-136.1328598}$
14. $\quad CSY_3 \quad = \quad Yc_2p(w_1c_{21} + w_2c_{22} + w_3c_{23}) \qquad = \qquad \underline{-125.4042161}$
15. $\quad CSZ_3 \quad = \quad Zc_2p(w_1c_{31} + w_2c_{32} + w_3c_{33}) \qquad = \qquad \underline{64.6256421}$

See E-32 for additional information on how the required values were obtained.

Example E-31: Two-Point Solution in Combined Geometry Imagery

The purpose of this exercise is to develop parametric values that can be used to support the three-point geometry of the trailer and sign (next to the warehouse and loading dock). In this example you will be using most of the methods in previous examples. The procedural steps are listed, references are made to the appropriate examples, and results are given. Consider this a test, and you must work through the example, correctly obtaining the same results. Note that subscript in this example refers to a two-point perspective solution value. Refer to Figure E-31, and Chapter Twenty-One, paragraph 21.6, for additional information. The photograph is cropped.

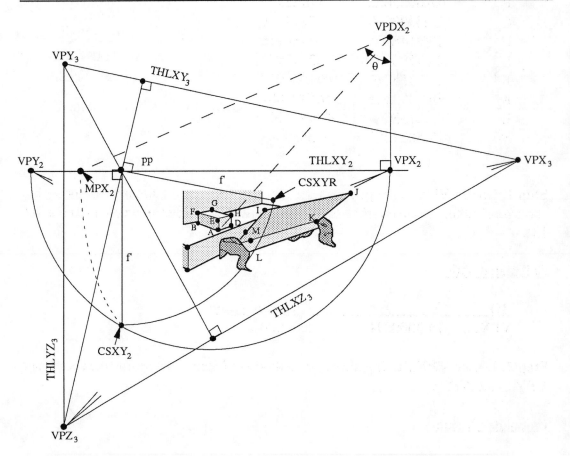

Figure E-31. Combined Two- and Three-Point Perspective Example

Required Values

Points are identified in Figure E-31

ID	x	y	(units = mm)
A	20.0000000	10.0000000	
B	18.6983662	10.5681734	
C	21.0000000	11.0000000	
D	22.3648453	10.4790668	
E	20.0000000	13.0000000	
F	18.6983662	13.4132175	
G	21.0000000	13.7272730	
H	22.3648605	13.3484116	
I	29.3796654	11.9393005	
J	37.2155876	13.2051039	
K	34.7094879	9.5408802	
L	27.9153099	8.0971174	
M	27.3823261	8.3369598	

Step 1. Locate VPX_2. Using three or more parallel lines, measure two points per line, and calculate the image coordinates $VPXx_2$ and $VPXy_2$ (see Chapter Thirteen and E-13).

Calculated Value

ID	x	y	(units = mm)
VPX2	74.3000031	21.0000000	

Step 2. Locate VPY_2. Using the same method as in Step 1, determine the coordinates $VPYx_2$ and $VPYy_2$.

Calculated Value

ID	x	y	(units = mm)
VPY2	-5.2000000	21.0000000	

Step 3. Transform the image coordinate system. VPY_2 is the new origin, and x-axis is coincident to THL_2 with +x is towards VPX_2.

1.	w_1	=	$VPXx_2 - VPYx_2$		=	79.500000
2.	w_2	=	$VPYy_2 - VPXy_2$		=	0.0000000
3.	w_3	=	$xA - VPYx_2$		=	25.2000000
4.	w_4	=	$yA - VPYy_2$		=	-11.0000000
5.	x_A	=	$(w_3*w_1 - w_4*w_2)/(w_1^2 + w_2^2)^{1/2}$		=	25.2000000
6.	y_A	=	$(w_3*w_2 + w_4*w_1)/(w_1^2 + w_2^2)^{1/2}$		=	-11.0000000

Repeat Steps 3 through 6 for points B through M. Then determine the coordinates of the midpoint C_2 between VPX_2 and VPY_2 on the THL_2.

Calculated Values

ID	x	y	(units = mm)
VPX_2	79.500000	0.000000	
VPY_2	0.000000	0.000000	
A	25.200000	-11.000000	
B	23.898366	-10.431827	
C	26.200000	-10.000000	
D	27.564845	-10.520933	
E	25.200000	-8.000000	
F	23.898366	-7.586782	
G	26.200000	-7.272727	
H	27.564860	-7.651588	
I	34.579665	-9.060699	
J	42.415587	-7.794896	
K	39.909488	-11.459120	
L	33.115310	-12.902883	
M	32.582326	-12.663040	

7.	C_2x	=	$VPXx_2 / 2.$	=	39.75000
8.	C_2y	=	$VPXy_2 / 2.$	=	0.000000

Step 4. Locate the vertical for **VPDVX**. The **VPDVX** will be located on line perpendicular to $THLXY_2$ and through VPX_2. (Remember, both sides of a 45° triangle are equal. Use **C** to provide a second point (P2) on the vertical line through VPX_2.)

Calculated Values

ID	x	y	(units = mm)
C_2	39.750000	0.000000	
P2	79.500000	39.750000	

Step 5. Locate **VPDVX**. Use the two-line intersection equations (E-13) to determine intersection of vertical through **VPX$_2$**, and diagonal of crate. This is the vanishing point **VPDVX**. If several diagonal angles (1 through k) are known, there can be k **VPDVXk** positions.

Calculated Values

ID	x	y	(units = mm)
VPDVX$_1$	79.500003	65.883980	
...	...	...	
VPDVXk	.	.	

Step 6. Locate **MPX$_2$** (CSXY2 rotated). With true angle (**TA**) calculate the **MPX$_2$** coordinates using MPXx = VPXx - (**VPDVX**y - **VPX**y) Tan(**TA**/Rad), and MPXy = 0 (origin is at **VPY$_2$**, and **THL$_2$** is the x-axis).

Calculated Value

ID	x	y	(units = mm)
MPX$_2$	6.4000000	0.0000000	

Step 7. Locate **CSXY$_2$**. Use **CSXY$_2$**-**C$_2$**-**VPX$_2$** and radius of the semicircle **VPX$_2$**-**VPY$_2$**. The side opposite **C$_2$** is **VPX$_2$**-**CSXY$_2$**, and equal to **VPX$_2$**-**MPX$_2$**. The angle (Ang1) at **VPX$_2$** is determined using the law of cosines. The cosine of Ang1 is cos(Ang1) = (**VPX$_2$**x-**MPX$_2$**x)/2(**VPX$_2$**x-**C$_2$**x). The sine of Ang1 is sin(Ang1) = (1.0-cos(Ang1)^2)1/2. The **CSXY** coordinates are **CSXY$_2$**x = **VPX$_2$**x - (**VPX$_2$**x-**MPX$_2$**x)Cos(Ang1), and **CSXY$_2$**y = -(**VPX$_2$**x-**MPX$_2$**x)Sin(Ang1).

Calculated Values

ID	x	y	(units = mm)
CSXY$_2$	12.284780	-28.735417	

Ang1 = 23.1473630^0

Step 8. Determine remaining Phase One parameters. The **pp**, **f'**, a_2, and s_2 can now be determined from the image coordinates of **VPX$_2$**, **VPY$_2$**, and **CSXY$_2$**. The tilt angle by definition is 90°. The swing angle is 180° as defined in Step-3. a_2 is found using right-triangle **VPY$_2$-THxy$_2$-CSXY$_2$**, and the equation is $a_2 = \tan^{-1}((\mathbf{CSXY}x-\mathbf{VPY}x) / \mathbf{CSXY}y)$. a_2 is also **Ang1** from Step 8. Equations to determine **[R]$_2$** are presented in Chapter Fourteen.

Calculated Values

ID	x	y	(units = mm)
pp	12.284780	0.0000000	

$$f' = 28.735417 \text{ mm}$$
$$a_2 = 23.147364°$$

$$[\mathbf{R}]_2 = \begin{bmatrix} .919496861 & -.393097345 & 0.000000000 \\ 0.000000000 & 0.000000000 & 1.000000000 \\ -.393097345 & -.9194968661 & 0.000000000 \end{bmatrix}$$

Step 9. Locate **XC$_2$**, **YC$_2$**, **ZC$_2$**. Use the procedures presented in E-23 or E-24. The value of **VPZx** will equal **ppx**, and the value of **VPZy** should be an extremely large number; e.g. $|1 \times 10^{+6}|$. It is recommended that you use the values of **VPZ$_2$** only as a means to check the results.

Required Values

ID	x	y	(units = mm)
A	25.200000	-11.000000	
B	23.898366	-10.431827	

DAB = 10.000000   feet

Calculated Values

ID	X	Y	Z	(units = feet)
CS$_2$	-136.1328598	-125.4042161	64.6256421	

Step 10. Determine object-space coordinates for points A, B, and H. This step takes us beyond the purpose of the exercise; however, there may be a need to know the object-space coordinates of points on the warehouse or loading dock. Use E-27 procedures to determine the two-point system object-space coordinates. Remember, this is a building block procedure.

Required Values

ID	x	y	(units = mm)
pp	12.284780	0.000000	
A	25.200000	-11.000000	
B	23.898366	-10.431827	
E	25.200000	-8.000000	

Required Values

ID	X (feet)	Y (feet)	Z (feet)
CS$_2$	-136.1328598	-125.4042161	64.6256421
A	0.000	0.000	0.000*
B	0.000	10.000000	0.000*
E	0.000*	0.000	17.625175

* Indicates the coordinates used as the known values in the calculations

Note: Do not map the coordinates to the objects illustrated in Figure E-31; the illustration is not to scale.

Example E-32: Three-Point Solution in the Combined Geometry Imagery

The purpose of this exercise is to have the dimensions and coordinates of the trailer and sign (E-31) in the same relative coordinate system as the warehouse and loading dock. The procedures for the three-point solution depend upon the solution of E-31. This is a continuing exercise. See Figure E-31, and Chapter Twenty-One for additional information.

Step 11. Locate VPX_3. Compute the image coordinates of VPX_3 using the same procedures described in Step 1.

Calculated Value

ID	x	y	(units = mm)
VPX_3	103.867366	2.131929	

Step 12. Locate VPZ_3. Using the same procedures of Step 1, determine the image coordinates of VPZ_3.

Calculated Value

ID	x	y	(units = mm)
VPZ_3	4.732655	-62.891914	

Step 13. Locate VPY_3. With the known image coordinates of the **pp**, VPX_3, and VPZ_3, compute VPY_3 using Procedure No. 3, paragraph 14.6. This procedure is very similar to E-12.

Required Values

ID	x	y
VPX_3	103.867366	2.131929
VPZ_3	4.732655	-62.891914
pp	12.284780	0.000000

Calculated Value

VPY_3	2.936846	14.251766

Step 14. Locate the **f'** using the procedures given in E-20.

Calculated Value
   f' = 28.735417

Step 15. Compute **[R]**$_3$. Calculate the rotation matrix elements using the procedures of E-19.

Calculated Values

$$
[\mathbf{R}]_3 = \begin{bmatrix} .953900435 & -.279795393 & -.108574892 \\ .022205623 & .426573254 & -.904180375 \\ -.299300635 & -.860087087 & -.413121468 \end{bmatrix}
$$

Step 16. Locate **XC, YC, ZC**. Use the procedures presented in E-23 or E-24. These are the same procedures used in Step 9.

Required Values

ID	x	y	(units = mm)
I	34.579665	-9.060699	
J	42.415587	-7.794896	

D$_{IJ}$ = 50.0 feet

Calculated Values

ID	X	Y	Z (units = feet)
CS$_3$	-121.1629215	-59.67757853	72.05735605

Step 17. Locate object-space points of I, J, K, L, and M, using the same procedures as in Step 10 (E-27 procedures). Remember, this is a building block procedure. Required Values

ID	x	y	(units = mm)
pp	12.284780	0.000000	
I	34.579665	-9.060699	
J	42.415587	-7.794896	
K	39.909488	-11.459120	
L	33.115310	-12.902883	
M	32.582326	-12.663040	

$f' = 28.735417$

ID	X (feet)	Y (feet)	Z (feet)
CS_3	-121.1629215	-59.67757853	72.05735605

$$[R]_3 = \begin{bmatrix} .953900435 & -.279795393 & -.108574892 \\ .022205623 & .426573254 & -.904180375 \\ -.299300635 & -.860087087 & -.413121468 \end{bmatrix}$$

Calculated Values

ID	X (feet)	Y (feet)	Z (feet)
I	0.000	0.000	0.000
J	50.000	0.000	0.000
K	50.000	0.000	-22.822
L	4.527	0.000	-22.822
M	0.000	0.000	-20.377

Step 18. Convert the object-space coordinates of CS_3, I, J, K, L, and M using the procedures of E-30.

Required Values

$$[R]_2 = \begin{bmatrix} .919496861 & -.393097345 & 0.000000000 \\ 0.000000000 & 0.000000000 & 1.000000000 \\ -.393097345 & -.9194968661 & 0.000000000 \end{bmatrix}$$

$$[R]_3 = \begin{bmatrix} .953900435 & -.279795393 & .108574892 \\ .022205623 & .426573254 & .904180375 \\ -.299300635 & -.860087087 & .413121468 \end{bmatrix}$$

Calculated Values

$$[C] = \begin{bmatrix} .994762741 & .358515951 & -.437602944 \\ -.099769734 & .782118405 & -.262209602 \\ .022205623 & .426573253 & .904180375 \end{bmatrix}$$

ID	X (feet)	Y (feet)	Z (feet)
CS$_3$	-136.133	-125.404	64.626
I	37.323	-71.924	27.620
J	87.062	-76.912	28.730
K	97.048	-70.928	8.095
L	51.814	-66.391	7.085
M	46.240	-66.581	9.196

NOTES

NOTES

NOTES

NOTES

NOTES

NOTES

NOTES